ALTERNATIVE PERSPECTIVES ON LIVELIHOODS, AGRICULTURE AND AIR POLLUTION

Dedicated to the village communities of Faridabad and Varanasi, who participated in this study and made it possible.

The present book is the outcome of participatory field research under the project *Community Perspectives on Impact of Air Pollution in Urban and Peri-Urban Agriculture.* The project was funded by DFID's Environment Research Programme, as part of a larger research project *The Impacts and Policy Implications of Air Pollution on Crop Yields in Developing Countries* of the T H Huxley School at Imperial College of Science Technology and Medicine in collaboration with International Institute for Environment and Development (IIED), London.

Alternative Perspectives on Livelihoods, Agriculture and Air Pollution

Agriculture in urban and peri-urban areas in a developing country

NEELA MUKHERJEE

with
MEERA JAYASWAL
BRATINDI JENA
AMITAVA MUKHERJEE
SUDIPTA RAY

Ashgate

Aldershot • Burlington USA • Singapore • Sydney

Published by
Ashgate Publishing Limited
Gower House
Croft Road
Aldershot
Hampshire GU11 3HR
England

Ashgate Publishing Company
131 Main Street
Burlington, VT 05401-5600 USA

Ashgate website: http://www.ashgate.com

British Library Cataloguing in Publication Data
Mukherjee, Neela
   Alternative perspectives on livelihoods, agriculture and
   air pollution : agriculture in urban and peri-urban areas
   in a developing country. - (SOAS studies in development
   geography)
   1.Urban agriculture - India - Varanasi 2.Urban agriculture
   - India - Faridabad 3. Air - Pollution - India - Varanasi
   4.Air - Pollution - India - Faridabad 5.Varanasi (India) -
   Social conditions 6.Faridabad (India) - Social conditions
   I.Title II.University of London. School of Oriental and
   African Studies
   338.1'0954

Library of Congress Control Number: 00-111840

ISBN 0 7546 1695 9

Typeset at dot design, New Delhi, India. *E-mail: dotdesign@vsnl.com*

Printed and bound in Great Britain by Antony Rowe Ltd., Chippenham, Wiltshire

Contents

List of Charts and Boxes

List of Tables

Preface

The present book "Alternative Perspectives on Livelihoods, Agriculture and Air Pollution, Agriculture in urban and peri-urban areas in a developing country" is the outcome of participatory field research under the project "Community Perspectives on Impact of Air Pollution in Urban and Peri-Urban Agriculture". The project was funded by DFID's Environment Research Programme, as part of a larger research project "The Impacts and Policy Implications of Air Pollution on Crop Yields in Developing Countries" of T H Huxley School at Imperial College of Science Technology and Medicine in collaboration with International Institute for Environment and Development (IIED), London. The project based on participatory field research has led to two volumes on the subject. The present book is the first volume based on synthesis of community perspectives from selected urban and peri-urban villages in Varanasi and Faridabad districts of the Indian States of Uttar Pradesh and Haryana, respectively. The second volume written by Amitava Mukherjee is entitled "Perspectives on Air Pollution, Crop Yield and Food Security in India, case studies from urban and peri-urban areas". This accompanying volume contains analytical case studies based on community perspectives from selected villages covered by the project.

The farmers' perspective is an alternate view on pollution, which is important from their position as primary stakeholders of development. Such views come from their direct field experience and need not necessarily match with the formal or the expert view on the subject. However, it is important to learn about farmers' perspectives for any realistic analysis and policy-making and for building micro-macro linkages. The views of the farmers, who are at the cutting edge of such air pollution have been neglected for too long with the scientists' views dominating the scene. This has resulted in farmers' 'worldview' getting marginalised in the process implying thereby underscoring of the social aspects of air pollution and over-emphasis of the scientific aspects of such a problem.

At present, though impact of air pollution on agriculture is a major concern of the farmers located in and around industrial and urban areas, very little has been written on such a topic. The research on agriculture and air pollution has been concentrated in terms of scientific experiments and expert opinion based on scientists/environmentalists' perspectives. Scientific studies show that air pollutants such as sulphur dioxide, fluorine, nitrogen oxide, ozone and particles of dust and soot emanating from urbanisation, industrialisation and vehicular traffic adversely affect agriculture in complex ways. Such air pollution has a major impact on productivity and efficiency of wo/man and nature and pose as serious health hazard leading to higher levels of morbidity, premature mortality and cases of chronic bronchitis and other respiratory infections.

So the issue is how does air pollution impact on lives and livelihoods of those farmers living in such polluted areas. Who can better explain such impact other than those farmers, whose livelihoods are directly dependent on agriculture bordering industrial areas. Since farmers' views are never looked upon with credibility and never given priority in conventional development practice there is a big gap between scientist's view from the 'top' and the grassroots' view from the 'bottom'. The present book and its accompanying volume try to bridge this gap by bringing a range of perspectives on the topic from the grassroots.

The present study has approached the topic of air pollution in a holistic manner in terms of local lives and livelihoods from the perspectives of local communities, in which impact of air pollution on agriculture has been viewed as part of a whole livelihood system. Field results show that many farmers, both women and men, at the periphery of industrialised areas are directly facing the negative externalities of polluted air and many are well aware of such complex impacts though not in terms of 'our' scientific jargons. With polluted air and also water, they are facing new problems in crop production for which they hardly have solutions. With rapid pace of industrialisation and urbanisation they are paying a price through reduction in crop yield, income, health and uncertainties in livelihood including loss of habitat through forcible acquisition of land by government housing agencies against low compensation. Some of them are coping with the situation by means of complementary livelihoods like raising livestock and selling milk in industrial areas while others are facing threatened livelihoods incapacitated by the low compensation for land acquired by the government agencies and absence of effective rehabilitation package. The local farming community covered by this study are under immense stress due to uncertainties that have cropped up in their agricultural practices and productivity; their conditions of living and well being; the stress on health; the land acquired by government in a high handed manner and the lack of state support. Their periodic representations to the government have proved ineffective and their 'voice' has fallen on deaf ears in the face of strong power lobbies for industrialisation.

In order to learn about farmers' perspective on impact of air pollution a team of participatory approaches' (PA) researchers conducted field research in urban and peri-urban areas of India at Varanasi (Uttar Pradesh) and Faridabad (Haryana) during August 1998 to November 1999. Such field research was carried out in selected urban and peri-urban areas of India, in Faridabad (14 villages) and Varanasi (14 villages). Going by scientists' criteria, only those villages were selected, which demonstrated different pollutant levels and were close to the sites where scientists were measuring air pollution levels and impact on crops. Such selection was refined and improved upon in consultation with village communities depending on different criteria such as location of village, social composition, size, level of development etc. The fieldwork was geared towards listening to primary stakeholders, in which around 1200 women and men farmers were listened to, of which about half were women.

The participatory field research was not limited by the topic of impact of air pollution on agriculture. It had a much broader scope covering lives and livelihoods of local communities, their agricultural practices, their health and well being, assessment of social cost/benefit of industrialisation and urbanisation including learning from local farmers' about the local impacts of air pollution. The objectives of the participatory field study were as follows:

- to learn from local women and men groups of farmers about the importance of agriculture and their livelihood strategies;
- to learn about the constraints in agricultural practices and their coping strategies;

- to learn about the impact of air pollution on agriculture, health and others;
- to learn about the doable's they suggest.

The field study was based on PRRA or Participatory Rapid Rural Appraisal, a variation of the methodology of PRA or Participatory Rural Appraisal, which is a widely used methodology for an interactive process of social development throughout the world. It is a way of learning from the people, with the people and by the people. PRRA is used for generating information required by external agents but expressed by the communities themselves in their own way and with their own emphasis (UNDP:1996). It is quite different from RRA or Rapid Rural Appraisal, which is quicker with information gathered from one or two members of a local community, which are then interpreted by external experts in their own way, in their own framework. The principle of PRRA is to acquire just enough information of poor people's own experience and analysis of the situation under consideration. Similar to RRA, it is generally a "one off" and extractive exercise but differs in its emphasis on the process of participation and ensures that people themselves define their priorities. It is important to recognize that PRRA is neither exhaustive nor conclusive. It is rather indicative and seeks to embrace the diversity of situations and diversity of people's own experience and perceptions of such situations.

For the field study, a team of researchers undertook PRRA for conducting field research. Individual researchers such as Meera Jayaswal, Bratindi Jena, Sudipta Ray, Hema Saklani, Sunita, Navin Datta-Banik and Madhumita and small teams from ActionAid, India, Council for Social Development, New Delhi and Indian Social Institute, New Delhi conducted field inquiry in the villages of Faridabad and Varanasi. Listening sessions were conducted with groups of women and men farmers and PRA 'visuals' techniques like participatory mapping, scoring/ranking, Venn diagramming and others were used. For learning about different perspectives gender was mainstreamed into the fieldwork and perspectives of old/young and rich/poor were also covered by the study. Amitava Mukherjee helped in analysing community perspectives from individual villages and also in comparing results from the two districts of Varanasi and Faridabad as in Chapter 4 of this book.

In order to share the field results, field workshops were organised and conducted in early 2000 with farmers in three villages, Sohtai, Sumper and Sahapur Kalan of Faridabad in collaboration with the agriculture extension department of Haryana Government. Six workshops were held separately for women and men farmers with the objective of flagging the major agricultural constraints identified during field inquiry. Such workshops proved to be a forum for farmers to raise issues with the extension department for clarification. For each session, the extension department officials tried to answer a volley of questions put to them by the farmers and also gave practical suggestions for over coming some of their farming constraints. The farmers found such workshops useful for clarifying issues and for building links with the extension department while the latter in turn thought that it was easy to approach the farmers through the medium of such workshops. However, only six workshops could be held in three villages of Faridabad within the limited budget. The response from the six workshops shows that such face-to-face communication workshops of farmers and officials hold considerable promise for opening communication channels given the fact that at present such functional relationships are either weak or non-existent.

The various results from the field study have been organised and synthesised for the present book. This book is divided into 6 chapters as described below.

Chapter 1 provides an introductory background to the study. It covers the research issues, the dimensions of field inquiry and a technical appendix on the problem of air pollution in urban and peri-urban areas.

Chapter 2 has a description of community perspectives from Varanasi on the role of agriculture, their farming practices and seasonal activity calendars, their agricultural problems and constraints, impact of air pollution on agriculture and health, effectiveness of their support system, community-based assessment of urbanisation and industrialisation and the "do-ables" at Varanasi. Annexure 'A' contains profiles of agricultural communities in urban and peri-urban areas of Varanasi.

Chapter 3 focuses on perspectives of agricultural communities from Faridabad on the role of agriculture, their farming practices and seasonal activity calendars, agricultural problems and constraints, impact of air pollution on agriculture, health and other social issues of such communities, effectiveness of their support system, assessment of urbanisation and industrialisation, and a description of the "do-ables" at

Faridabad. Annexure 'B' contains profiles of agricultural communities in urban and peri-urban areas of Faridabad.

Chapter 4 provides a comparative picture of livelihood and agriculture at Varanasi and Faridabad with regard to constraints in agriculture and coping strategies, perceived impact of air pollution on agriculture, institutional issues, health problems confronting the two communities and a comparative summary of the quality of life.

Chapter 5 draws upon the lessons learnt from the field and discusses them under generic clusters. Such generic clusters relate to approaching community for participatory interactions, listening and learning from them, conducting field inquiry, making research culturally compatible, applying PRA methods, maintaining quality and pace of research, handling sensitive issues, managing community expectations and communicating with scientists and other stakeholders.

In Chapter 6 the methodology of field inquiry has been described including the ways adopted for conducting both exploratory and topical rounds of field research. Criteria for village selection has also been discussed and different participatory methods used by the researchers have been listed. The field research cycle has been explained in order to provide contextual meaning to the kind of results arrived at. The chapter also includes a generic checklist of issues probed and the code of conduct of the research team.

In this book, the field results indicate important inter-relationships between livelihood, agriculture and air pollution as analysed by men and women farmers. In the villages studied, agriculture both directly and indirectly, was found to be the most important source of livelihood where 50 per cent to 90 per cent of the villagers were dependent on agricultural activities including livestock. In villages of Varanasi (Uttar Pradesh), the main crops were paddy and wheat, with land under wheat considerably less than land under paddy. In case of villages in Faridabad (Haryana), wheat was the principal crop, grown for consumption while jowar for fodder with vegetables grown mainly as cash crop. Livestock rearing and milk vending were very important activities for the Faridabad farmers with jowar being used as fodder. Though the intensity of the agriculture constraints varied, the constraints, in general were related to weather problems, weed-attacks, pest attacks, animal attacks, pollution, crop disease, problems of land, inadequate infrastructure, shortage of credit and weak institutions. The emphasis of farmer groups was on the collapse of the state support system both in Faridabad and Varanasi, low access to agricultural inputs and no surveillance of unscrupulous traders cheating farmers.

With regard to air pollution, the farmers of Faridabad villages said that air-pollution was linked to industrial belts in near-by towns of Ballàvgarh and Faridabad, which were close to the villages. The smoke and different pollutants from local factories and industrial units like those of brick kilns, thermocol, plastic and cement pipe factories affected crops yield and health of both human beings and animals. Some women and men farmers in Varanasi villages described the sources of local pollution as industrial estate, industrial units, domestic coal stoves, garbage and vehicular pollution. On the whole, the sources and impact of air pollution were perceived more in the villages of Faridabad than in the villages of Varanasi. Amongst the "do-able's" suggested by the farmers a major one was offsetting damages from industrialisation and urbanisation process by strengthening agricultural policy support system and formulating policies that promoted sustainable agriculture. The field researchers were of the view that community awareness building on impacts of air pollution and ways of mitigating them were immediately required as policy interventions.

The book aims at sharing field results with different professional groups associated with issues related to livelihood, agriculture and air pollution. This book and its accompanying volume on case studies are meant for development practitioners, environmentalists, development agencies, donor agencies, government departments of agriculture and urban/rural development and pollution control boards, NGO's in urban/rural development working in health, environment, development and gender issues, agricultural scientists, agriculture economists, social activists, policy makers, extension agents, pollution abatement specialists, urban housing agencies, pollution control authority, green bench of judiciary, public interest groups, field researchers, students of urban/rural environmental planning and those working in the area of corporate governance and corporate ethics.

It is my privilege to mention that there are many individuals, who have made the field research and writing of this book possible by contributing from time to time. I am thankful to DFID's Environment Research Programme for the funding and The T H Huxley School at Imperial College of Science Technology and Medicine, Ascot, Berks, UK and the International Institute for Environment and Development (IIED), London, UK for the opportunity provided for undertaking the present study.

Our research team owes an intellectual debt to the village communities of Varanasi and Faridabad in urban and peri-urban areas who were instrumental in making the field research so rich and interesting. They provided their scarce time and space for discussion and took great interest in sharing their experience and knowledge with the research team. We express our sincere thanks to all the participants in the villages of Faridabad and Varanasi without whose support and cooperation the present study would not have been possible.

I thankfully acknowledge the invaluable support and encouragement provided by Dr. Fiona Marshall from The T H Huxley School at Imperial College of Science Technology and Medicine, Ascot, Berks, UK. Such periodic support was critical for bridging gaps between the scientist's perspectives and the perspectives of farming communities from the field.

I express my sincere thanks to Simon Croxton of International Institute for Environment and Development, London for providing a stimulating environment for new research issues and the flexibility in sorting them out. I wish to put on record my appreciation for Dolf te Lintelo's administrative support and his comments on an earlier draft of this study. Thanks are also due to John Thompson of International Institute of Environment and Development who initially invited me to undertake this research work.

It is important to mention the contributions of Dr. Madhulika Aggarwal of Benaras Hindu University, Varanasi in her unstinted support for the field work at Varanasi. It was with her that consultations were held about the kind of field results and the available scientific evidence, which helped in clarification of many issues. The National Institute for Urban Affairs, New Delhi especially Dr. Madhushree Mazumdar was also of much help at the start of the project. It was at that Institute the first two meetings were held with the Participatory Approaches research team.

Thanks are also due to India International Centre, Delhi where the first peer-group seminar and other meetings were held. Services of the library of the India International Centre and the British Council, New Delhi, were also of much help. Thanks are also due to Sandip Ahuja for helping with the camera-ready copy. My sincere thanks are due to Ashgate Publishers, who have published this book and its accompanying volume of case studies in collaboration with Development Tracks Research Training and Consultancy, New Delhi.

All recommendations and suggestions to further this field study are welcome. There lies considerable scope for undertaking similar studies in different geographical areas thus providing voice to the voiceless for flagging issues in livelihood, health and pollution due to the unsustainable kind of development taking place through unregulated growth of industries. It is important that those local communities, which are stressed by the impacts of industrialisation and urbanisation find opportunities to voice their grievances and make their rightful claims for offsetting damages done to their livelihoods, health and quality of life. I hope this book provides useful ideas to those working in the areas of farming communities, pollution and other related issues. I am solely responsible for the errors and omissions in this book.

The message of the book is loud and clear. Rampant pace of industrialisation and urbanisation and their externalities/impacts are playing havoc with the lives and livelihoods of countless farmers by weakening agriculture systems and human and animal health and increasing poverty, disease and uncertainties. In order to learn about such impacts it is important to listen to the farmers, both women and men about their views and experience. These can then provide gateways to building micro-macro linkages in policies and actions towards sustainable development, a roadmap for arresting and mitigating the impacts of industrialisation and urbanisation and for strengthening agriculture support systems and local livelihoods.

Neela Mukherjee
Coordinator,
Project: "Community Perspectives on Impact of Air Pollution",
New Delhi, India

Glossary of Indian Terms

Terms	Meaning
Aganwadi	A government child welfare day care centre where children up to the age of 5 receive free food and health check up.
Baithak	Living room.
Bajra	A coarse cereal.
Banga	Disease afflicting buffaloes. It affects the feet and it gets swollen.
Baroo	A type of weed which looks similar to jowar and is found mostly in jowar fields.
Barsam	Fodder sown in winter.
Basti	Ward or colony in a village organised mainly on the basis of caste/religion.
Bathua	A weed.
Bigha	Measure of Land: 5 Bigha = 1 acre.
Biswa	Measure of Land: 5 Biswa = 1 acre.
Burela	Another name for caterpillars.

Chepa	Tiny white insects — suck juice out of leaves of cabbage, bajra, mustard and sugarcane leaves. Has a sticky texture.
Chhedak	Any caterpillar that bores holes. It is also called Dhola/Burela.
Chitli	A worm that eats vegetable roots.
Chula	Oven or stove made out of mud which uses wood, animal dung etc. as fuel.
Congress Grass	A weed having small white flowers. Also called Parthenium.
Dhencha	Medium sized fodder plants sown in summer.
Dhola	Caterpillars that bore holes.
Fatinga	Any pest which is a grasshopper.
Galagotu	Disease found in buffaloes. It chokes buffaloes to death.
Gander	A general term for caterpillars.
Gram Samiti	A village co-operative society.
Gram Sewak	An appointee of the Block Development Office for supervising development work in the village. He/she looks after 3 villages.
Gurchawa	A disease which results in wilting and crippling of leaves.

Hara Kira	Small green caterpillar.
Harwawa	A disease which turns leaves yellow.
Jhansi	A cluster of small caterpillars (pest).
Jhulsa	A disease resulting in wilting and blackening of leaves.
Jowar	An inferior cereal used as fodder now.
Kanduwa	A disease which affects the ear-rings of wheat and turns them black.
Kasba	A village which is the centre of rural growth having a population of 7000 or more.
Kateli	A medium sized tree-weed having thorns.
Katui	A pest which cuts down paddy ear-rings.
Kirona	A caterpillar.
Koria	Crippling of leaves and plants like a leper. The fruits of the affected plants look deformed.
Lalrii	Red coloured small sized pests which can fly. It eats leaves of leafy vegetables.
Machchi/Makhi	House-fly.
Mandi	Wholesale market.

Morya or Gurchawa	A disease which wilts and cripples leaves especially in chilies and tomatoes.
Muhalla	Wards/Colony.
Najla	Ailment: running nose, headache, nose burning.
Nakshatra	A constellation.
Nilagai	Blue bull.
Pankhi	Any pest which can fly.
Pradhan/Sarpanch	Village head, usually elected to the village level unit of governance.
Rani Jai (Ghan Jai)	A weed similar in looks to wheat and whose seeds are like wheat grains, though slightly black in colour.
Safed Moondi	A weed having small seeds (coriander type), mainly found in wheat fields.
Sambar	A kind of deer.
Sati	Small sized weeds which mainly grow in bajra fields.
Soondi	Small sized caterpillar found in paddy leaves. It sucks the sap.
Telchatta	Small sized pest which can fly. It sucks juice out of mustard pods and cauliflower.

Tidda	A current variant of locust which is small in size.
Tiddi	Large sized locust (the earlier variant).
Tikuli	A beetle that sucks juice from leaves of vegetables mostly found in summers.
Titri	A small white moth which sucks milk from paddy ear-rings.

Months of the Year in the Indian Calendar

Vernacular Month	Approximate Corresponding Period in Roman Calendar
Baisakh	Mid-April to Mid-May
Jeth	Mid-May to Mid-June
Asard	Mid-June to Mid-July
Sawan	Mid-July to Mid-August
Bhado	Mid-August to Mid-September
Kuar*	Mid-September to Mid-October
Kartick	Mid-October to Mid-November
Aghan	Mid-November to Mid-December
Poos	Mid-December to Mid-January
Magh	Mid-January to Mid-February
Fagun	Mid-February to Mid-March
Chait**	Mid-March to Mid-April

* In some villages called Aswin.
** Also spelt as Cheth.

1 Introduction

Agriculture in urban and peri-urban areas of Haryana and Uttar Pradesh plays an important role not only as a source of livelihood but also in supplying food grains, vegetables, pulses, milk, fruits and other agricultural produce to urban and peri-urban communities and markets. It constitutes a major source of income and livelihood both for those who own land in and around such areas and for those offering their labour for employment. The range of food grains, pulses, cereals, vegetables and milk, produced by farmers in such areas are in part for self-consumption and in part for sale in the market. Such produce from agriculture helps in not only supplying essential commodities in the food chain but also in stabilising prices and markets and in maintaining stocks.

Issues, Questions or Problems Investigated

The present study was undertaken to learn about the relationship between agriculture in urban and peri-urban areas and the phenomena of industrialisation/urbanisation from the *farmers' perspective*, both women and men. Some related issues are the role of agriculture, the nature of agricultural constraints, the impact of air pollution on health status and quality of life and the support systems for addressing the problems of the farmers. There is enough room for widely divergent perspectives of both the relationships between pollution and damage to agriculture and the risks involved. Farmers, it has been widely recorded, have their own perspectives for a variety of reasons. Such perspective is important for realistic analysis and policy-making; ownership and/or maintenance of corrective interventions and for building micro-macro linkages in policy issues. The scope of the present research is learning from local communities about impacts of industrial pollution on agriculture, especially in urban and peri-urban areas. The community perspective is an alternate view on pollution, which is important from the point of view of their position as stakeholders. Such views come from direct experience of such impact and need not necessarily match with the formal or the expert view on the subject. It is in this context that the present research focuses on community perspectives for deriving appropriate policy measures and follow up action, for effective management of such impact.

A series of fieldwork based on farmers' perspective was carried out in selected urban and peri-urban areas during the period August 1998 to November 1999. The fieldwork was geared towards listening to primary stakeholders - women and men farmers in urban and peri-urban areas about their agricultural activities, practices and constraints. The sphere of agriculture broadly included crop and vegetable cultivation and livestock raising. Field investigation also included the impact of agricultural constraints on agricultural production, livelihood and health with focus on environmental problems arising from industrialisation and urbanisation. The field research explored the support system for agriculture, the role of local institutions in order to study their effectiveness and farmers' perspective on suitable ways for capacity building of such institutions.

Some of the issues addressed in the research study are as follows.
- Role of agriculture and constraints faced by farmers.
- How important is urban and peri-urban agriculture to them?
- Understanding the issues of air pollution from local communities' perspective.
- How significant is the impact of air pollution in agriculture?
- How does it compare with other agricultural constraints?
- What policy issues emerge from such community perspectives?
- How effective are the local institutions and support system for the farmers?
- What is the health status of local communities and what is its relationship with industrialisation and urbanisation?
- What more can be done in this regard?

Field Inquiry

Field research was undertaken at two sites in India, Faridabad and Varanasi. Micro level information was generated on the research topic through participatory rural appraisal (PRA) methods in selected agro-based villages of Faridabad and Varanasi located near industrial areas. A Participatory Approaches (PA) Team undertook such field research periodically during the crop year 1998-99. Altogether 14 villages of Faridabad and 14 villages of Varanasi were covered, selection of which was primarily based on scientific sites with different types of air pollution. The names of the Faridabad villages are Kadhaoli, Sagarpur Uncha Gaon, Baroli, Pali Kasba, Chandawali, Jajru, Piyala, Sahupura, Sahapur Kalan, Sumper, Sohtai, Malerna and Jharsainthly. The names of those villages covered in Varanasi are Chandpur, Navampura Kala, Seer Govardhanpur, Sarai Dongre, Lohta, Chitaipur, Nathupur, Tikri, Maraon, Maheshpur, Adityanagar, Karamanbir, Nuaon and Tarapur.

The field inquiry was based on community interactions using Participatory Rural Appraisal (PRA) methods (Mukherjee, 1995, UNDP: 1996). The villages for community interactions were selected based on sites where the scientists were experimenting on this impact under the project. Such selection of villages was refined and amended in consultation with village communities. The researchers carried out participatory field study as per the project objectives. Gender was mainstreamed into the fieldwork for approaching local community and learning about different perspectives (e.g. those of men/woman, old/young, rich/poor) in the communities undertaken for the study.

The present study provides a synthesis picture of the villages covered under field study at Varanasi and Faridabad. In this context, it is important to remember that open-ended participatory inquiry can generate asymmetric data depending on community perspectives and priorities. In the present study, field inquiry based on generic checklist of issues has been relatively open-ended and hence, by default space has been created for considerable data asymmetry in the research topics considered for the study. However, this has not been a hindrance in evolving broad contours for arriving at meso level pictures of Faridabad and Varanasi. Rather it has been an opportunity to flag community perspectives in the way that they have been voiced and also reflect their relative priorities as perceived by different village groups. The volume accompanying this report contains case studies on selected villages of Varanasi and Faridabad.

References

Mukherjee, Neela (1995 reprint), *Participatory Rural Appraisal, Methodology and Applications,* New-Delhi, Concept Publishing Company.
UNDP (1996), *Report on Human Development in Bangladesh,* A Pro-Poor Agenda, volume 3, United Nations Development Programme, Dhaka.

Technical Note

The phenomena of industrialisation and urbanisation currently sweeping the Indian economy have both positive and negative externalities for such agriculture. For one, they increase demand for goods produced in urban and peri-urban agriculture, expand markets, and increase employment and income opportunities; for the other, they adversely affect agricultural production, alternate livelihoods, health and quality of life, including social capital. Most often, industrial and urban growth pose major threats to agriculture by polluting natural environment and spoiling social environment. Rise in levels of air and water pollutants [1], are characteristic features of industrialisation and urbanisation, often accentuated by non-implementation of anti-pollution laws (UNDP, 1992). Air pollutants that are most damaging to agriculture are Sulphur Dioxide and the Oxides of Nitrogen, which are categorised as acid pollutants, and Ozone together with other photochemical Oxidants. Fluorides are another set of pollutants.

Air pollution in the urban and peri-urban areas comes from both urbanisation and industrialisation. It is singularly important to distinguish the challenge facing almost all urban and peri-urban areas in India from these two sources. Urban and peri-urban areas are either located along roads which, carry vehicular traffic or are visited by motor vehicles for a wide variety of reasons. Nitrogen oxide, released by the burning of fuel in vehicles, being an unstable compound, instantly combines with Oxygen to form Nitrogen Dioxide. This, in sunlight, gives out an extra molecule of Oxygen to form Ozone. The SPM provides a convenient platform for chemical reactions, which creates a haze like effect, seen mostly in the morning. Ozone is concentrated in the morning because Nitrogen Oxide reacts with the previous night's Ozone to start this chain chemical reaction. By noon when vehicular emission is high, Ozone starts forming and some of it escapes because of a rise in temperature. As evening descends and temperature falls, the earth becomes a piston, compressing the pollutants. The Ozone starts to drift to areas with lower density. The peri-urban areas are the receptor sites where even if the people are not culprits in contributing to the "smog", they receive the brunt of it. When Ozone drifts to peri-urban areas it destroys crops in large quantities. The first signs of this are seen in the leaves of the trees which shrivel up. This is extremely dangerous for farming households as well for those who are exposed to the outdoors for long hours and who are engaged in hard physical labour. On an average, a human being inhales 10,000 to 20,000 litres of air but those engaged in hard physical labour like farmers, inhale upto 70,000 litres of air, increasing the risks of exposure to smog.

Industrialisation produce, inter alia, Sulphur Dioxide (SO_2) by burning fossil fuel containing Sulphur. We are not aware of any data regarding India, but generally 70 per cent of Sulphur emission comes from power generating stations. And many urban and peri-urban areas are in the vicinity of power plants.

It is also well established that the highest ambient concentration of Sulphur Dioxide occurs in and around cities and near centres of Industrial activity. The adoption of efficient dispersion mechanisms may reduce the severity of SO_2 pollution close to its sources but this may be at the expense of increased ambient levels over much larger areas. However, in India as in the case of other developing countries, high levels of SO_2 are not so widely dispersed and they tend to be concentrated in and around urban areas and are usually localised. For instance, ICRISAT at Hyderabad produces potentially phytotoxic levels of several hundred parts per billion of SO_2, occasionally peaking 1000 parts per billion yet portions of the research station site, only 1-3 km from the sources experience background levels of only 5 parts per billion of SO_2 (Collins and Harris, 1983).

The effect of Sulphur Dioxide can lead to visible injury, which can take the form of necrosis of plant tissue—typically the leaves die from the tip of the plant downwards. It can also lead to growth reduction in certain plants like radish and tomato and cause fall in yields of barley (Conway et.al., 1995).

Fluorine is the other major pollutant. Flourine is typically emitted as hydrogen fluorine from brick kilns and tile works, steel works, potteries and fertiliser factories. Effect of fluorine in reducing yield has been noticed in several countries in wheat, onion, potatoes and barley (Halbwachs, 1984). Hydrogen fluoride also interferes with pollen germination and pollen tube growth. Fluoride in nectar and pollen can be passed to bees and other pollinators, susceptible to fluoride poisoning. And there is no natural mechanism for detoxification.

A major problem for agriculture is the presence of particles, mainly dust and soot in the air. Coal burning produces very high levels of suspended particulates. Today, deposition of dust from cement works, motorway construction, power stations and quarries and from traffic on motorways affect agricultural produce. Usually, the loss to the farmer arises because high value produce, such as fruit, vegetables and horticultural crops, are rendered unsaleable (Conway and Pretty, 1995).

Though the individual pollutants can cause damage to agriculture, polluted air usually consists of a mixture of pollutants, which may adversely affect agriculture in a more complex way. SO_2 pollution effects become confounded, for instance, by the presence of Nitrogen Oxides. In such situations several types of interactive effects may occur, termed as synergistic (Roberts, 1984), antagonistic, predisposition and desensitisation (Runeckles, 1984).

Air pollution has a major impact on productivity and efficiency of wo/ man and nature and pose as a serious health hazard. The most talked about impact is the health impact of air pollution leading to higher levels of morbidity, premature mortality and cases of chronic bronchitis and other respiratory infections.

References

Collins, J. J. and D. Harris (1983), 'Air Pollution assessment at the international Crops Research Institute for the Semi-arid Tropics, India', *Environmental Technological Letters*, Vol.8, 1983, pp.10785-10787.

Conway, Gordon and Jules Pretty (1995), *Unwelcome Harvest*, London, Earthscan and New-Delhi: Vikas.

Halbwachs, G (1984), 'Organisational responses of higher plants to atmospheric pollutants: sulphur dioxide and fluoride', in M. Treshow (ed.), *Air Pollution and Plant Life*, Chichester, John Wiley and Sons.

Khoshoo, T.N. (1984), 'Environmental Concerns and Strategies', India Environmental Society, New Delhi.

Markhan, Adams (1994), *A Brief History of Pollution*, Earthscan Publications Ltd., London.

National Institute of Urban Affairs, (2000), 'The role of urban and peri-urban agriculture in metropolitan city management in the developing countries', Research Study Series, Number 74, New Delhi.

Roberts, T.M. (1984), 'Long Term Effects of Sulphur Dioxide on Crops: An Analysis of Dose-response Relations', Phil. Trans. Royal Society London, No.305, 1984.

Rodhe, H. and R. Harrera (1988), *Acidification in Tropical Countries*, Chister, John Wiley and Sons.

Runeckles, V.C. (1984), 'Impact of Air Pollutant combinations on plants', in M. Treshow (ed.), *Air Pollution and Plant Life*, Chichester, John Wiley and Sons.

Simmons, I.G. (1993), *Interpreting Nature, Cultural Constructions of the Environment*, Routledge, London.

UNDP (1992), *Human Development Report*, New Delhi, Oxford University Press.

World Resources Institute (1998-99), *World Resources*, A Guide to the global Environment, Oxford.

Endnote

We are only considering air pollution here. Hence the subsequent discussion does not make any reference to other forms of pollution for e.g. water pollution.

2 Farmers' Perspectives from Varanasi: An Underdeveloped District in an Underdeveloped State

Introduction

Chapter 2 is divided into nine sections, where Section 2.1 relates to perspectives of agricultural communities from Varanasi on the role of agriculture. Section 2.2 is on farming practices and seasonal activity calendar of such agricultural communities. Section 2.3 relates to their agricultural problems and constraints and their priority issues. Section 2.4 describes their perspectives on impact of air pollution on agriculture. Section 2.5 contains health and other social issues of such communities, while section 2.6 outlines effectiveness of their support system. Section 2.7 is a community-based assessment of urbanisation and industrialisation. Section 2.8 provides a description of the "do-ables" at Varanasi. Annexure 'A' contains profiles of agricultural communities in urban and peri-urban areas of Varanasi. Selected PRA charts on Varanasi are given in Appendix 1 and the locational map of villages in Appendix 2.

Varanasi—Background

The district of Varanasi is located in the eastern part of the state of Uttar Pradesh, India. It covers an area of 403501 square kilometers with a total population of 3782949. Agriculture and horticulture are sources of major livelihood for the district of Varanasi. It also has an industrial base including small and medium scale industries. Many of the industries are agro-based, where the city of Varanasi with a wide transportation network, provides important wholesale markets for agro-based goods, handicrafts and industrial products. The engineering industries are mostly medium scale industries. There is a range of items produced by the small-scale industries as well, such as weaved Banarasi sarees, embroidery work and different wooden toys. It is also well known for its handicrafts, silk fabrics, beads, jewelry etc.

The district of Varanasi is flushed by several river systems by the Ganges, Gomti, Karmnasa, Chandraprabha and Varuna and is rich in natural resources. There are four sub-divisions in Varanasi—Varanasi, Chakia, Chandauli and Sakaldiha. There are three blocks in each of the sub-divisions with 1260 villages in Varanasi sub-division, 503 villages in Chakia, 428 villages in Chandauli and 436 in Sakaldiha. The city of Varanasi is an international tourist spot and is a religious hub for the Hindus and Buddhists.

All the villages under the present study were selected from the Kashi Vidyapith block of Varanasi sub-division, based on sites selected for the scientific component of the project. The Kashi Vidyapith block has about 129 villages covered by 13 Panchayats. In the block there are 15331 cultivators and 5993 agricultural labourers and 496 persons engaged in livestock and forestry. Around 42.1 per cent of total working population of Kashi Vidyapith block are directly engaged in agriculture and related activities. Of the rest, nearly 37 per cent are engaged in household industry, other industries and business and commerce, a major proportion of which is agro-based and indirectly related to agriculture. Hence, agriculture plays an important role in socio-economic life of Varanasi.

2.1 Farming Communities at Varanasi

At Varanasi, field research was undertaken in 14 villages, selection of which were based on scientific sites with different types of air pollution. The names of the villages are as follows.

• Adityanagar
• Chandpur

- Chitaipur
- Karamanbir
- Lohta
- Maheshpur
- Maraon
- Nathupur
- Navampura Kalan
- Nuaon
- Sarai Dongre
- Seer Govardhanpur
- Tarapur
- Tikri

Farmers' Perspectives on the Role of Agriculture—Varanasi

As described by the local communities from villages covered during field research at Varanasi, most households are primarily engaged in agriculture, which forms the backbone of their livelihoods. Whether agriculture gets practised in a village or residents of one village go to another for providing agricultural labour or for share cropping, the crux of the matter is that agriculture is of prime importance to local communities at Varanasi, in the urban and peri-urban areas. When discussing about agriculture the women and men farmers naturally include crop and vegetable cultivation and livestock. As expressed by them, the present role of agriculture is not only in terms of livelihood, employment and income generation but also in terms of critical support that it provides towards food and livelihood security on a seasonal basis for overcoming seasonal poverty. Agriculture provides a source of multiple livelihoods, both direct and indirect employment, asset base, consumption, income, savings, investment, collateral, source of risk coverage, raw materials and markets to the villagers covered under the study. As described by the local groups, brief profile of agriculture in each village is provided below in box 2.1.

Box 2.1
Agriculture as a Livelihood, Varanasi

- **Tikri**—In this village, located near the city, agriculture is the primary occupation of 90 per cent of the population, in which both young and old are engaged. Apart from agriculture some educated youth are employed in the local university while many work as wage labourer. Among the landed class of Tikri, some persons from older generation (very few) are engaged in agriculture, while others have rented their land and are doing business or service.

- **Maraon**—In this village, 75 per cent of the population is engaged in agriculture and for the past 7 years, the Government has initiated land consolidation in the village. Livestock also plays an important role in the livelihood of the local community with a high goat population.

- **Maheshpur**—Agriculture has a low priority in Maheshpur, where only 4 to 5 acres are left for agricultural purposes with 20-25 households engaged in such activity in "Chowri Bazar" locality.

- **Chandpur**—In the village of Chandpur, main livelihood activities consist of agriculture, livestock, service and wage labour. Role of agriculture is limited since the proportion of cultivable land has declined and people have diversified their livelihood from agriculture to services in both private and government (around 7 per cent of total villagers are currently engaged in agriculture).

- **Navampura Kala**—Role of agriculture is considerably limited in the hamlets of village Navampura Kala.

- **Nathupur**—Agriculture is one major source of livelihood in which three crops are grown including a variety of cereals, pulses, vegetable and oil seeds and fruits like guava.

- **Seer Govardhanpur**—Agriculture and livestock constitute 60 per cent of livelihood, while wages make for 30 per cent and trade 10 per cent of the total. Share of agriculture in total family income is 66 per cent. Agriculture is not spread throughout the year.

- **Lohta**—Weaving forms a pre-dominant occupation and involves the muslim households. Nearly 30 per cent of households are involved in agriculture while 70 per cent are rural artisans. Agriculture makes for 25 per cent of total household income.

- **Chitaipur**—The village has 90 per cent of households dependent on wages and agriculture and 10 per cent on livestock. However, in terms of contribution to household income, the contribution of agriculture is only 40 per cent.
- **Sarai Dongri**—In this village, livelihood activities are diverse with 35 per cent in agriculture, 20 per cent in livestock, 38 per cent as wage labourers, primarily agriculture, 5 per cent in services and 1 per cent, each in trade and rural artisans.
- **Adityanagar**—Adityanagar, spread over 12 acres of land, has no agricultural land. Majority of men folks are engaged in activities related to beads making and machine embroidery and women make garlands of beads and roll bidis. A few are engaged as agricultural labourers, cultivators and share-croppers in adjacent villages. The village has around 250 households of which approximately 180 are wage labourers in different activities including agriculture. Some of the households cultivate vegetables in their kitchen gardens.
- **Karamanbir**—In village Karamanbir, about 50 per cent of the total area is agricultural land, of which, 75 per cent is low level land growing paddy and wheat while 25 per cent is high level land growing vegetables throughout the year. The residents of this village are mostly farmers and agricultural labourers. Average land holdings are of small size, though a few own more than 7 bighas. In order to increase returns from agriculture, households are growing more vegetables.
- **Nuaon**—Agriculture is being practised as a major livelihood. Village Nuaon has a total land of 400 bighas and distribution of land for agriculture is very uneven. At least 60 per cent of land is own by Bhumiars whose main livelihood is agriculture, while 90 per cent of Patels, 50 per cent of Dalits and Rajbhar households have low land holdings and work as agricultural labourers.
- **Tarapur**—The village is spread over a large area with 170 households where most are engaged in agricultural activities either as land owners or as agricultural labourers. The women groups other than Bhumiars and Brahmins also work as agricultural labourers. Agriculture is practised in three seasons, summer, rains and winter.

Source: Based on Field Reports of PA researchers from Varanasi

The villages covered during field research at Varanasi exhibit different profiles in agriculture. Agriculture, in general, a major occupation is perceived differently by local communities. Box 2.2 provides a glimpse of such perceptions from selected villages.

Box 2.2

**Perspectives on Agriculture from Urban and
Peri-Urban Areas in Varanasi**

- **Village Karamanbir**—Villagers of all hamlets depend on agriculture either as cultivators or as labourers. With low level of literacy, the local people are still able to contribute to agriculture because of their having farming skills. The women group is of the view that agriculture supports daily expenditure by 50 per cent for majority of poor groups.
- **Village Nuaon**—Majority of people, both women and men, from the four hamlets of Nuaon village are associated with agriculture. The larger the land holding, the larger is the benefit from agriculture. Agricultural land not only satisfies immediate needs but also provides opportunity for getting loan during crisis. Because of such intrinsic value of agricultural land the women would like to retain such land despite its high market value. They want to pass such land to their posterity. Even the landless aspire to purchase agricultural land.
- **Village Tarapur**—Women group perceive that one of the crucial benefits of agriculture is its potential to meet their food requirements. Without one's own produce it is not possible to have unadulterated food. Agriculture is a form of disguised employment for majority of the population, both landowners and landless. The women can also actively contribute in agricultural activities. It provides significant support during emergency. Hence it is not advisable to sell land, which is the best gift to transfer to one's sons.
- **Village Adityanagar**—In Adityanagar, one women group pointed out that agriculture is a fixed deposit, which can be encashed during crisis. There is also the possibility of using land as collateral for getting loans. According to the group, agriculture offers continuous employment and also high returns in case of vegetable cultivation. Since agriculture protects against crisis, without agricultural base the no. of crisis periods would be on the rise. The women group is of the opinion that it is possible to eat fresh and unadulterated food only from one's own farm.

In terms of income, agricultural activities in the villages, under consideration, serve the dual purpose of household consumption and main/supplementary income (such as in case of vegetables). For instance, in the villages under study, households mostly consume the cereals grown, whereas, vegetables are largely marketed.

Livelihood and Coping Strategies at Varanasi

Patterns of livelihood were analysed by the local communities at Varanasi. In most villages, studied during field research, agriculture is important in terms of household livelihood, though its contribution to family life and income varies across villages. The villagers are engaged in a range of activities, many of which are directly and indirectly related to agriculture. Some of the livelihood-related activities for income generating purposes are shown in box 2.3.

Box 2.3
Livelihood-related Activities in Villages of Varanasi

- farming
- selling agricultural labour
- share cropping
- raising livestock
- selling milk
- working in service sector
- bead making
- garland/necklace making
- tractor lending
- shop keeping
- electric repairing
- masonry work
- working in power loom units
- TV repairing
- auto rickshaw driving
- vegetable vending
- weaving
- tailoring, hosiery stitching
- others

Livelihood, Agriculture and Gender Issues

- The village women (mostly those belonging to lower caste as in Hindu caste hierarchy) are fully engaged in agriculture and livestock-related activities. In village Maraon, the women, in addition, carry and sell vegetables in the markets.
- The women in Maraon and Tikri (mostly from middle and lower income groups) supplement their family income by making "bidi" and bead necklaces.
- Apart from household work and raising livestock, some women in village Chandpur and Navampura Kala sell their labour in agriculture, few others are engaged in making necklace of "Rudraksh" seeds while other are engaged in biri (made of *tendu* leaves) making.
- The women group/s at Varanasi described some of the agriculture-related work done by them as listed in box 2.4.

Box 2.4
Work done by Women in Varanasi Villages

- call tractor for field
- sow seeds
- irrigate field
- remove weeds
- thresh and pound paddy etc.
- prepare rows for sowing crop
- sprinkle fertilisers
- apply medicines/pesticides to the field
- look after and oversee crops
- harvest crops
- sell vegetables
- other related work

Source: Field Reports from Varanasi of PA Researchers

Many of the men folks are engaged in providing agricultural labour and other labour, trading/ business like running tea shops, selling betel leaves, cycle repairing shops etc. and providing labour to factories, construction sites and saree printing units.

Shifts in Patterns of Livelihood

In terms of employment, the trend shows an occupational shift from agriculture to services though there is considerable dependence of villagers on agriculture. The farming households constitute mostly of marginal and small farmers thus not having much scope for generating wage employment. The household trend in procuring food is also gradually shifting from self-production and consumption within households to growing dependence on outside markets. It is also perceived that there is an overall decline in share of agricultural activities in favour of manufacturing/service sector. The proximate factors for the declining role of agriculture are, at least, three.

- one, constraints and risks in agriculture are on the rise thus affecting returns;
- two, the manufacturing and the service sectors are providing jobs; and
- three, agricultural land is being disposed of for non-agricultural purposes.

The village women and men groups pointed out that the service sector has absorbed a proportion of wage labour. Some have also started their own enterprises like opening outfits/shops for selling different commodities.

Seasonality of Livelihood and Quality of Life

Monsoon is the worst period, when the daily labourer finds it difficult to get jobs. Also rain makes village roads muddy and constrains mobility on a daily basis. See box 2.5.

Box 2.5
The Crisis Month of 'Bhado' (mid August-mid September)

Why is there crisis in the month of Bhado?—as explained by women group of Karamanbir:

- The stored grains like paddy and wheat are consumed.
- Daily earners do not get work due to excess rain.
- It is difficult to move during heavy rain.
- Agricultural land is green with paddy and vegetable plants but there is no income.
- It is difficult to arrange loans from acquaintances for all suffer from similar conditions.
- Due to crisis the poor are the most vulnerable and they can afford just one meal a day.

Source: Field Report on Karamanbir, Varanasi by Meera Jayaswal

2.2 Farming Practices and Seasonal Activity Calendar

As described by the local women and men groups, the seasonal calendar in table 2.1 show that the farmers from peri-urban areas of Varanasi grow different kinds of rabi and kharif crops. They pursue agricultural activities throughout the year. Their rabi crops include main crops like wheat and others vegetables like raddish, cauliflower, chana/peas, potato/onion/garlic, spinach, methi etc. Amongst kharif crops are included paddy (main crop), maize, jowar, arhar, cauliflower and brinjal. Multiple cropping is done for e.g., in the Rabi season (from Aghan to Baisakh) with early cultivation of different vegetables for quick income followed by cultivation of wheat mainly for personal consumption.

- In most of the villages, crops like wheat, paddy, jowar, maize etc. are grown mainly for home consumption whereas, the vegetables grown are mostly sold in the market.

- Over the years, changes in cropping pattern has led to cultivation of 2-3 cereal crops with paddy and wheat occupying dominant position. The cropping pattern has changed from millet to fine cereals like paddy, wheat and intensive cultivation of vegetables due to some improvement in irrigation facilities, mostly on a private basis.

- Earlier, the land holdings in the villages under reference were large and had higher production but, in recent years, the size of land holdings have become considerably small due to sub-division and fragmentation of land and sale/use of land for construction purposes. With diminished land size, vegetables are being substituted in place of crops.

- The use of wage labour for agricultural activities is minimal and is limited to ploughing of field and other activities with general wages ranging between Rs.30/- Rs.35/- per day.

- As regards food security dimensions in poor households, agricultural production can meet the cereal requirements to some extent. For other items of consumption, the villagers depend on the market. A small proportion of households in the villages is also engaged in sharecropping and shares half the produce.

- With slight increase in irrigation facility, the farmers are growing more paddy and wheat rather than millets though such irrigation facility is not enough. With some improvement in irrigation facilities, wheat crop is being substituted in place of jowar, bajra and makka.

- Soil in the villages under study has become heavily dependent on fertilisers, consequently the cultivators find it difficult to grow barley and gram. Farming system practised in the villages reflect semi-modern techniques though farmers have adopted improved variety of seeds.

- With reduced land size, around 90 per cent of the farmers are growing vegetables to realise good profit. Such reduced land size has made it inconvenient to grow cereals in both the seasons. Though, in case of Navampur Kala, land holdings are bigger in size as compared to earlier times due to consolidation of land holdings and change in ownership of land.

- In some villages such as Tikri and Maraon, cost of cultivation is generally covered by income from other sources. In village Maraon, litigation expenses for on going court cases have reduced the share of agriculture in the last two years.

Table 2.1   Seasonal Calendar: Activities Related to Agriculture— Varanasi

Jeth-Asar (15 May-15 July)

In Jeth, paddy is sown; in Asar, paddy seedlings are planted; loan is taken for agricultural purposes; in Asar, spinach is sold; harvesting and threshing of wheat is done; some onions from the field are harvested, suplus over and above personal consumption are sold in the market; jimikand is sown for personal consumption; in Asar, turmeric is sown for personal use; maize and paddy seeds are used for personal consumption; some ninwa, lady's finger and moong are sown; tomato is also sown, of which, popular seed variety ganga gaurav and pusa red are prefererd since they bear fruits early (but in the current year, tomato plants died due to rains); jimikand is harvested and fields cleared; pigeon pea and bajra are sown for fodder use and for personal consumption.

Sawan (15 July-15 August)

Vegetable, maize and paddy are cultivated; paddy saplings are planted; cabbage seeds are sown; in Sawan, carrot is sown; brinjal, tomato, chilli, spinach, lady's finger, bora etc are also sown.; farmer makes a choice between cultivating spinach or cauliflower; the former has least cost and can be cut 3-4 times; cauliflower cultivation needs more funds, though returns are high; along with the above soya is sown; spinach, if sold without soya fetches less price; weeding is also done in paddy fields; maize, moong and jowar are also sown.

Bhado (15 August- 15 September)

Weeding is done in paddy fields followed by application of aluminium sulphate and zinc sulphate for better growth; paddy saplings are planted; some women work as agricultural labourers, do weeding; look after the crops sown in Asar; bajra is sown; mustard is sown for personal use; maize is harvested; in Bhado, loan is taken for agricultural purposes either from fertiliser shops or from lenders in the village; Kuari cauliflower is also sown (some sow it on an elevated plain so as to protect the crop from excess rain).

Kuwar-Kartick (15 September-15 November)
Income accrues from agriculture; fields are ploughed before sowing of potatoes; carrot and raddish are sown; pea, gram and potato are sown (if surplus output it is kept in cold storage); garlic, coriander, methi and spinach are sown for personal use; bajra and paddy are harvested; Kuari cauliflower is sold; income accrues by selling spinach from first round of cutting; potato is sown; double cropping is generally practised with vegetables sown earlier in the season and wheat crop follows; main crops are wheat and cauliflower; moong is harvested in Kuwar; vegetables are harvested and sold; maize and jowar are also harvested; in Kartick, fields are ploughed for next season.

Aghan-Pus-Magh-Phagun (15 November to 15 March)
Harvesting of paddy and threshing of paddy takes place; harvesting of pigeon pea is also done; late wheat is sown; fields with wheat sown earlier are irrigated; harvesting of mustard takes place; pea is sown; field is made ready for potato sowing; potato is sown by mid-Aghan; wheat is sown in Aghan last; before which, the field is ploughed several times with fertilisers; potato seeds are purchased; in mid-Aghan carrot are often sown in potato fields; prior to sowing, field get ploughed; before sowing fertiliser, potash etc. are applied to the termite prone areas; in Aghan, seasonal cauliflower is sown whose leaves are smaller than the earlier seasonal variety but their flowers are comparatively large; water is sprinkled to plants; peas, gram and sugarcane, onion, potato, cauliflower, brinjal, tomato, carrots are sown by Aghan-end; tilling of potato fields and second round of irrigation takes place; in Pus, carrot fields are irrigated when plants are big; pea fields are irrigated when they start bearing fruits; second round of irrigation takes place for wheat crop; urea is applied to wheat; weeding is done in all fields if the weather is good; lady's fingers, chillies and tomatoes, those sown in Bhado are harvested; these and other vegetables are harvested, vegetables are washed to remove the soil and the best ones are picked for sale; cauliflower, spinach, raddish and other vegetables are carried in basket over head to the market for sale; cauliflower is sown in Bhado/Aghan, spinach and raddish are sown in Kartick, cauliflower is sold in Bhado; in Pus wheat is sown, mainly Malwi 234/253; variety; every second year seeds are changed because the yield tends to fall; irrigation is done for early sown wheat; irrigation and tilling are done in potato fields and plants provided with soil cover; urea is sprayed in wheat fields after irrigation; income starts accruing from agriculture; in Magh/Phagun, carrots are sold, ninwa is sown; wheat is sown, mainly Malwi 253 variety; in Magh, mustard is harvested; weeding of fields takes place; sugarcane is sown; fertilisers are applied to carrort fields when they are slightly wet; pea fields are harvested, the surplus is sold in the

market and fields cleared; in fagun, early sown wheat is harvested; in Phagun, garlic, coriander; methi, onion, peas and gram fields are harvested; ninwa, bitter gourd, lady's finger, cucumber, kakri, gourd, small creeper melons and sugarcane are sown; in Phagun there is some time for rest and time to look after livestock; potatoes are harvested from the field; peas are sold.

Chet-Baisakh (15 March-15 May)
Mainly spinach and ninwa are sown in Baisakh; gourd, raddish, kakri, pumpkin, coriander, cucumber and kohra are also sown; main crops are spinach (Punjab palak) and ninwa; tomatoes are harvested; nitrogen is sprinkled in the fields of ninwa, bitter gourd and lady's finger; wheat is harvested; threshing of wheat begins; winter vegetables are harvested; gram fields, mustard and pigeon pea fields are harvested and cleared; tilling takes place in onion fields; 2-3 rounds of irrigation are done; less income accrues from agriculture; ninwa is harvested and the surplus sold in the market.

Note: This is a general seasonal calendar as described by women groups and men groups from the urban and peri-urban villages of Varanasi covered during field study. Not all households of the villages are engaged in all activities mentioned above. There are different crops and vegetables sown in different villages and the above table provides a general picture of agricultural activities undertaken in the villages under study. For village-specific activity see field reports related to the villages under study

Source: Based on field reports of PA researchers on Varanasi

In most villages under study, annual return from cultivating paddy and wheat has relatively low rate of return as compared to growing vegetables or selling milk. With regard to wage payment, daily wage received by men labourer is higher as compared to that of women while salary from providing services is the highest. In Navampura Kala, livestock as a source of livelihood is more lucrative than agriculture/cultivation. In Maraon villages, around 75 per cent population are engaged in Banarasi saree weaving, a popular income earning activity due to its proximity to Lohta village of weavers.

2.3 Agricultural Problems/Constraints

Local Communities' Analysis of Problems/Constraints

There is a range of problems/constraints in agriculture as described by women and men groups of the villages under reference. Such agricultural problems/constraints can be clustered into seven generic groups. A glimpse of village-specific problems are given in box 2.6 as described below.

Box 2.6
Difficulties of Farmers: Tikri Village, Varanasi

- Cauliflower saplings died this year at least 4-5 times because of unseasonal rains.
- For the last 8-10 years damage of crops by Nilgai and stray cows have become real problem.
- Incidence of "Jhulsa" disease has increased in the last 15 years in brinjal, tomato and chilly affecting 90 per cent of such vegetables.
- There is the constraint of rising input prices in agriculture.
- When the crop fails the farmer is not able to save seeds and is forced to buy seeds from the cooperative (against loans).
- If there is any compensation from the government against crop losses due to natural causes such as hailstorm, the landowner gets the benefit instead of the cultivator.

Source: Field report on Tikri, Varanasi, Sudipta Ray

All these generic groups of problems/constraints are important to the farmers, though, as we shall see below, their priorities differ. The emphasis and the dimensions of the problems tend to differ across locality. For example, with regard to infrastructure/input-related problems, one agriculture community emphasizes lack of irrigation facility while another emphasizes supply of adulterated pesticides/insecticides. We thought that it was important to explore the local nature of each generic problem and the significance of each issue. The generic problem and its local dimensions have been assembled in box 2.7.

Box 2.7
Issues within Problem-Clusters in Agriculture, Varanasi

• **Weather Condition-related**—problem of hailstorm, frost, wind and fog; problem of recurring flood; rainfall not enough; crops getting damaged due to weather fluctuations; excess rain; moist wind due to rain; unseasonal rain.

• **Weed-related**—damage by genhu ka mama.

• **Pest and insect-related**—problem of too many pests; pest attack at the flowering stage; pest attack in paddy; attack by different pests like katui, tikuli, bahadur kira, macchi (pest fly), bhurila (pest hairy caterpillar) , phanga; chedak; dhoka; maho etc.

• **Animal-related problem**—crop damage by nilgai, rats and stray cows; nilgai grazes on crops.

• **Disease and Pollution-related problem**—fields damaged by polluted water from factory; decrease in level of fruit production; incidence of agricultural disease increased in the last 4 years; lack of treatment; yellowing of wheat crop; crop disease in arhar and paddy; lemon orchard not bearing fruits for many years; red dust accumulation over wheat leaves and low crop yield; yellowing of mustard leaves and low yield; smoke as pollution; crop diseases like safra, hardwa, gurchawa, jhulsa, oktwa, tunki, angarwa etc.

• **Infrastructure/Input-related problem**—small size of cultivable land; cultivable land far from home; lack of water for fields; inadequate irrigation facility; high price of irrigation; lands lying fallow due to lack of irrigation; shortage of water for wheat; lack of tube well; forced to fetch drinking water from a distance; no tube well installed by the government; reservoir of water getting contaminated; constraint of land ceiling act; constraint of scattered land; lack of animals/tractors for ploughing; shortage of electricity; seeds and fertilisers are expensive; supply of fertilisers is of low quality; supply of adulterated fertilisers; supply of adulterated pesticides; good quality seeds not available; high price of fertilisers and seeds; insecticides/ seeds not available in time and forced to buy them at higher price; problem of getting insecticide and seeds at right price; costs of potato and onion seeds were very high; high cost of share cropping; improper road hence difficulty in transporting fruits/vegetables; loan/credit not available in time; acute shortage of credit and cash; complex process of getting loans; no proper institutional facility for getting loan; high cost of loan; shortage of wage labourers; heavy workload; lack of knowledge of scientific cultivation; lack of knowledge of proper selection and use of seed; no scope for agriculture-related training; problem of adolescent boys.

* **Organization-related problem**—communication gap between villagers and officials; local cooperative centre closed down due to corruption and bankruptcy; officials related to village institutions corrupt; no service delivery of Kisan sewa bank; prices of farmers' produce not decided by farmers.

Note: As described by women groups and men groups from selected villages of Varanasi

Source: Based on field reports of PA researchers on Varanasi

Seasonality of Agriculture Problems/Constraints

One important dimension of agriculture problems/constraints at Varanasi is that they are seasonal in nature. The seasonality aspects are indicated in table 2.2. The table provides an idea of those seasons when the problems/constraints become acute and when interventions for minimising or overcoming the problems can be planned. One season when the pest-related problems are high is that of the rainy season Sawan-Bhado, when a large number of pests and insects cause damage to the crops as shown in table 2.2. Crop diseases are prevalent throughout the year while infrastructure-related problems are more in the summer months.

Table 2.2   Agriculture Problems/Constraints/Risks in Varanasi

Jeth-Asar (15 May-15 July)

Weather-related—unseasonal excesss rain damage flowers in tomatoes.

Weed-related—genhu ka mama which competes with wheat was more for the last three to four years and its growth was under control; ghour jai (weed) found in the wheat fields though no trend pattern found; same is with bathua (weed) in the wheat fields.

Pest and Insect-related—more damage due to termites.

Disease-related—hardwa (disease) is found in wheat though no trend reported; kanduwa disease in wheat due to which ear rings turn black, no trend reported; jhulsa disease in onions, no measure; gurchawa in tomatoes, increased incidence in the last 3 to 4 years due to adverse weather conditions; this impacting on yield and reducing size of the tomatoes; banjha in tomatoes; fruit borer in tomatoes due to which 1 or 2 plants died in every 20 plants; jhulsa in tomatoes; abnormal increase in height of 1 or 2 tomato plants in 20 plants, noticed earlier too.

Infrastructure/Input-related—due to shortage of irrigation facilities vegetables cannot be sown; over the last 5-6 years, the problem is that the generator, if used for threshing the wheat cannot be used for running the pump sets.

Sawan-Bhado (15 July-15 September)

Weather-related—rain can damage spinach; weeding the field during heavy rain becomes difficult; during rain, water has to be drained out of the vegetable fields to save the plants.

Pest and Insect-related—pests like phatinga, tiddi etc. attack the crops; khukri (pest) in jowar; dhola (pest) in the flowers and fruit of lady's finger, which is increasing for the last 2-3 years leading to considerable loss; dhola (caterpillars) in ninwa; fruit borer attacks; Makki (fly) attacks the ear rings of young paddy and sucks the milk out of it (this occurs more in late sown paddy); katul attacks the paddy, when ear rings have grains in them, tears the ear ring and pulls it down; in all seasons, cauliflower is attacked by Jhansi (black diamond moth) and the flower disfigures; sainik keet is active during night at paddy, grows fast, attacks when paddy is ripening, eating into the ear rings of paddy; termites found in the field; damage in vegetables due to grass hoppers.

Disease-related—hardwa disease attacks the paddy due to the weakness of the soil (at youth stage) (due to excess water); in all seasons, size of cauliflower vary widely with 4 kinds of size, small, medium, big and bigger, share of each around 25 per cent of total crop; gurchawa disease takes place in tomatoes; some tomato plants grow tall but don't bear fruits; this year for the first time, tomato crop did not germinate; jhulsa disease in paddy, during fruiting time, when leaves turn yellow and dry; katui disease in paddy.

Animal-related—damage due to nilgai.

Infrastructure/Input-related—This is the lean season where expenditure on agriculture is higher than income; limited sowing for many due to lack of irrigation facilities.

Kuwar-Katak/Kartick (15 September-15 November)

Pest-related—hara kira pest in sem.

Disease-related—In Kuwar, the normal yield of cauliflower is 700-800 per bigha but currently is only 600 per bigha; bajra ear rings can get diseased in which the ear rings turn slightly black and grains may not grow; jhulsa disease in spinach seen for the last 15 years when loss reported is 90 per cent; charkhawat (disease) in sem when the plant is around 4 feet, it takes place when heat is more; in kuwar, kurra disease in bajra during fruiting time leads to wilting of leaves; these are seen for the last 3 to 4 years.

Aghan-Pus-Magh-Phagun (15 November-15 March)

Weather-related—The crop gets damaged when fog is present; last year fog damaged leaves of tomatoes, potatoes and peas; the crop gets damaged with rain and hailstorms; at the time of ripening of wheat ear rings if there is rain and blowing of strong winds then grains become thin and less (resulting in 50 per cent loss); if it rains then vegetable crop gets damaged; frost and fog mainly damage peas, pigeon-pea etc., due to the cold the farmers find it difficult to go to the field and if it rains it is still difficult; large loss of potatoes due to last year's fog where potato leaves get blackened and fall; frost and hail affect potato crop; if no rain in winter then 100 per cent loss of crops.

Weed-related—due to genhu ka mama (weed), the damage in wheat crops can be to the extent of 2 per cent; genhu ka mam (weed) affects wheat production (loss can be 2 kg per bigha).

Pest-related—flowers of cauliflower get attacked by pakhi and other pests for which spraying of insecticides prove ineffective; pests occur in potatoes.

Disease-related—flowers of cauliflower get attacked by pakhi and other pests for which spraying of insectides prove ineffective; In Phagun, during excess rain or adverse weather there is hardwa disease in wheat, when the ear rings are coming out; the hardwa disease adversely impacts the size of grains, which become thin, small and less in number; gurchawa in brinjal with 75 per cent loss reported; banjha in brinjal with 500-600 plants, only 1 or 2 cases; uktwa disease in peas, plants dry bottom with up to 50 per cent loss; safra (disease) in peas when fruits have come, plant turns white and yield falls with 30 per cent loss; jhulsa disease in onion when about to bear fruit, mostly in February; udraha disease in chilli where its branch falls on its own; some ear ring turn black from the side; jhulsa disease in wheat seen for the last 15-20 years affecting yield marginally.

Animal-related—stray cows destroy crops; rabbits eating grams for the last 2 to 3 years; loss due to nilgai.

Infrastructure/Input-related—expenses rise due to sowing.

Chet-Baisakh (15 March-15 May)

Pest-related—pests like tikuli, bhaluwala keet (hairy caterpillar) damages the vegetables; jhansi pest damages kakri and cucumber saplings; heat makes it difficult to do field work.

Disease-related—Gurchawa disease attacks the ninwas when about to fruit; similar problem of wheat as in Phagun; jhulsa disease in tomatoes, whose incidence has increased in the last 4-5 years due to more heat/dryness in summer; gurchawa and chitkawra affecting ninwa and gourd; gurchawa occurs because of untimely rain and easterly wing affecting bitter gourd, its incidence has increased over the 2 to 3 years.

Animal-related—last 2-3 years, rabbit eating bitter gourd and ninwa.

Infrastructure/Input-related—due to water shortage vegetables cannot be cultivated; whatever is sown generally dries due to irrigation.

Note: As described by women groups and men groups from selected villages of Varanasi
Source: Based on field reports of PA researchers on Varanasi

Prioritised Problem Index of Local Communities, Varanasi

The next issue is whether all problems/constraints in agriculture are equally important or whether some are more important than others. The issue is about the way in which the problem/constraint clusters were prioritised by the women and men group/s of Varanasi villages. The listing, scoring and ranking of problems/constraints in agriculture as done by different women and men group/s have been assembled for inter-community comparisons based on an elementary statistical tool of indexing. Some groups used fixed scoring (i.e. expressing priority through apportioning a fixed number of seeds/stones to the problems), other used free scoring (using any number of stones/seeds/etc. to score), others only ranked problems without scoring and some groups listed their problems without scoring or listing. Hence, there emerged a range of ways in which groups did their problem prioritisation. For comparison across communities in Varanasi, the scores, ranks and frequencies of problem listing have been indexed to arrive at prioritised problem indices (PPI) of agricultural constraints/problems (see technical note). This has helped in retaining diversity of problems and maximum information, which local communities from Varanasi have provided. Such PPIs have been constructed for women and men groups separately as shown in table 2.3 and in bar chart 2.1.

Table 2.3   Prioritised Problem Indices (PPIs) of Agricultural Constraints/Problems—Varanasi

Problem/Constraint Criteria	Rank by Women	Rank by Men
Weather Conditions-related	5	5
Weed-related	7	6
Pest and Insect-related	3	2
Animal-related	4	4
Disease and Pollution-related	2	3
Infrastructure/Input-related	1	1
Organization-related	6	7

Note: The above is based on the prioritised problem index of the village community resulting from listing, scoring and ranking exercises done by local groups of women and men in Varanasi villages. The various groups managed the process of prioritising problems differently, while some used problem ranking, others did problem scoring and still others merely listed their problems. For problem scoring, many groups used seeds or stones to indicate their priorities (with greater the number of seeds greater the priority)

Source: Based on field reports of PA researchers on Varanasi

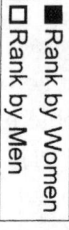

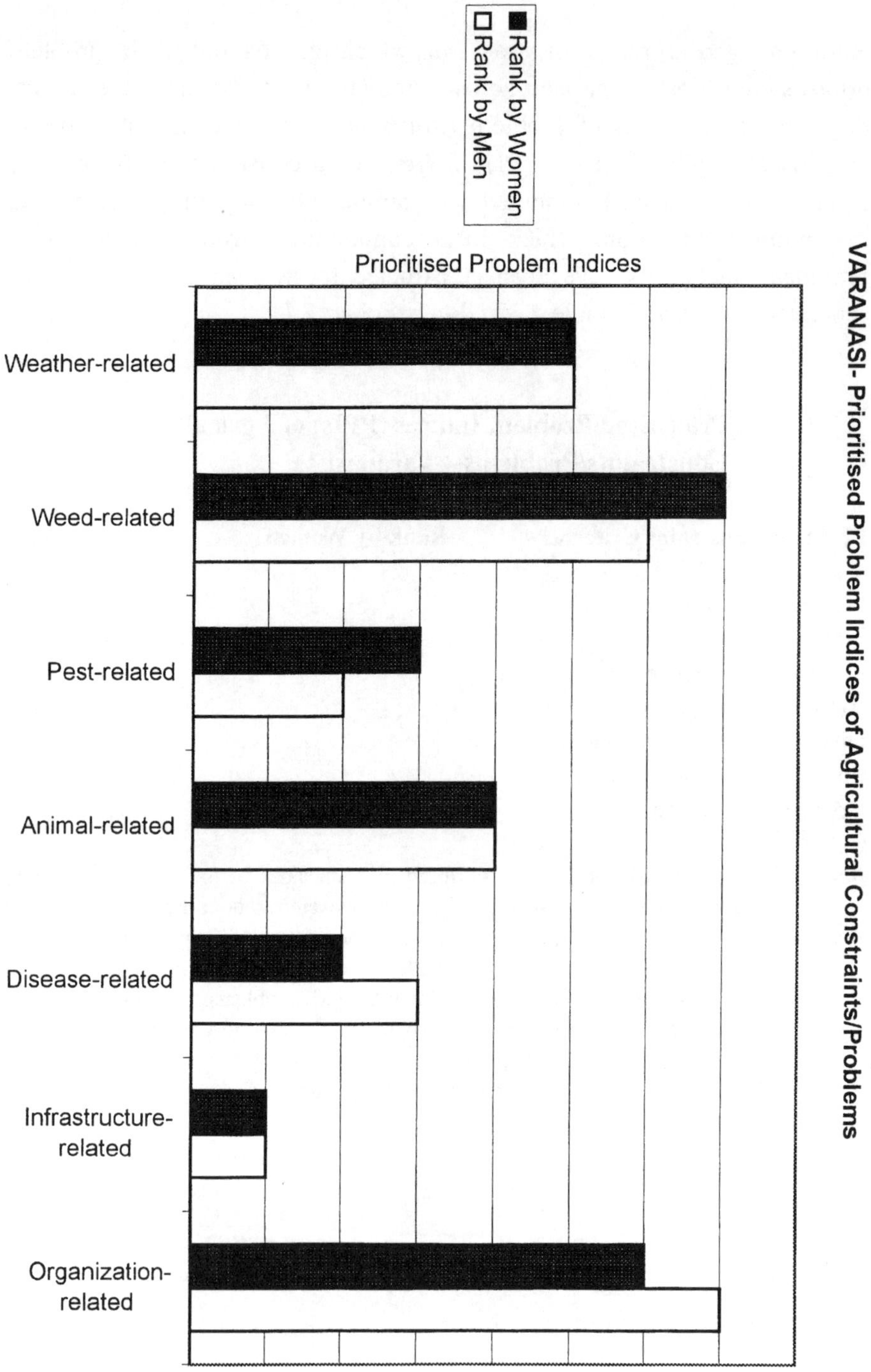
VARANASI- Prioritised Problem Indices of Agricultural Constraints/Problems
Prioritised Problem Indices
Rank by Women
Rank by Men
Weather-related
Weed-related
Pest-related
Animal-related
Disease-related
Infrastructure-related
Organization-related

The PPIs in table 2.3 and chart 2.1 show that while prioritizing agricultural constraints/problems, views of both women and men groups of Varanasi villages are found to converge in identifying infrastructure/input-related set of problems as the priority set of problems. This is because they find it difficult to cope with, especially the bureaucratic inefficiency for example of the extension department and the lackadaisical attitude of the state support system. Such problems have intensified over the years, adversely impacting production and livelihoods, eating into the returns and adding to cost, both direct and indirect. See boxes 2.8 and 2.9. Problems regarding finding the right kind of agricultural inputs at the right time and at the right price add to the risk of the farmers. Lack of supportive infrastructure like irrigation facilities, cheap credit sources, and absence of proper road for marketing and rising input costs make agriculture a difficult proposition.

Box 2.8
Crop Diseases and Pest Attacks—Villages Chandpur and
Navampur Kala, Varanasi

In the last 5 years, there has been an increase in crop/vegetable-related diseases and illness. The incidence of such diseases and pests are more during the ripening stage of the crop/vegetable and hence, adversely affects its production to the extent of 40 to 60 per cent. The pesticides prove ineffective for crops but work on vegetables. The farmers think that such pest attacks are due to erratic weather and untimely variations in micro-climatic conditions, which affects soil conditions. With more houses under construction the flow of water without outlets gets collected in the fields and increases dampness and pests.

Earlier, with application of organic fertiliser there were less pest attacks and they were tackled through sprinkling of ash. Earlier, tomatoes were planted for commercial use on 1 to 1.5 bigha/s but due to pest attack since 5 to 6 years the farmers have stopped growing tomatoes. Earlier, harvest of jackfruit was good but now the harvest has reduced by 20 per cent. Earlier there was excellent production from pumpkin plants giving large size pumpkins. But now, though the plant grows it does not bear any fruit.

Source: Field Report, village Chandpur and Navampur Kala, Varanasi by Hema Joshi and Sunita Bisht

---

Box 2.9

Problem of Plant Disease—Village Karamanbir, Varanasi

"While discussing plant diaseases, the women group of Bichalapura hamlet, village Karamanbir, requested me to see the diseased plants and insects. I saw damaged brinjal, its shape deformed and also the colour. The leaves of lady's finger were pale, the flowers and vegetables being damaged. The damage was also seen in tomato. One of the women pointed out at the red-black coloured insect, which was damaging leaves, flowers and tender vegetables."

Source: Field Report, village Karamanbir, Varanasi by Meera Jayaswal

---

The second prioritised problem of women groups relate to crop disease and the adverse impact of pollution. Problem of crop disease is third in priority for men. Farmers have to cope with a variety of crop diseases, some of which are relatively new while others, which are prevalent since long years. Though the farmers are aware of some treatment for certain crop diseases, many of the diseases cannot be cured or reduced through treatment. Some plant diseases have increased over the last few years and afflict the plants at different stages of crop cycle, from germination stage to harvesting.

The situation is aggravated by the problem of pollution, which manifests in different ways. Often the fruit production falls/stops, fields are damaged by polluted water from near-by factories, leaves get yellow, leaves curl up, crop yield gets affected by dust, smoke etc. and the diseases get aggravated in presence of bad weather.

The third problem, prioritised by women, is that of pest, of which there is a wide variety affecting the crop at different stages of growth and the available treatment through pesticide is not that effective. Men disagree with women and put pest as the second prioritised problem. The farmers, both women and men, find the pest problem more important than that of nilgai or blue bull, which is the fourth problem for both men and women groups. The nilgai eats the crop and inflicts loss, though wheat is saved from nilgai since they find it difficult to eat. The other crops are vulnerable to the attacks from nilgai, which comes in herd of 10 to 15 and cause damage to standing crops. Change in weather and its conditions though adversely affecting agricultural production are relatively less important to both women and men who place it as the fifth constraint.

Weed-related problems are low on priority as shown in the table 2.2, which is ranked by women as sixth while men rank it as seventh because regular weeding of crops and vegetables helps in neutralising impact of many weeds though not that easy to tackle.

The men rank organizational constraint as the sixth one, while women rank it as the seventh. The weaknesses of the existing organizational structure and linkages such as,

- Inefficiency and bankruptcy of the Kisan Sewa bank.
- Corruption of government officials at the implementation level.
- No empowerment of farmers to decide farm price.
- Bankruptcy of cooperatives.
- Lack of scientific knowledge of cultivation and pests/disease.
- Lack of communication between villagers and officials,
- And all others that add to the constraints and problems posed for sustaining agriculture as a livelihood.

Women group/s rank weed-related problems as the sixth while men groups rank it as the seventh constraint.

Illustrations of Some Village-specific Constraints/Problems

Some village-specific constraints/problems are listed in box 2.10 for having an idea of the problem/constraint at the local level.

Box 2.10
Village-specific Constraints, Varanasi

Weather Condition-related

- In Navampura, almost every year there is the problem of flood as a result of which kharif crops do not get planted.
- In Chandpur and Tikri, the villagers perceived the change in weather, which is getting increasingly hotter and there is untimely rain. This is affecting almost all food grains and vegetables.
- Villagers of both Tikri and Maraon perceived change in weather as one major factor behind loss of yield. It was also perceived to be the reason catalyzing pest attack.
- Some from Tikri pointed out man made satellites as one of the factors causing change of weather. Some from Maraon said that as per Hindu calendar next year would have 13 months instead of 12. This will rectify the weather change and consequent pest attack.

Weed-related

- In Tikri and Maraon, weeds weren't considered to a serious problem because weeding was done from time to time. Increase in crop disease was attributed to changing weather conditions.

Pest and Insect-related

- In Navampura, over the years, the problem of soil getting infested with termite has increased.
- In Maraon, major agricultural constraint is that of increased pest attack.
- The incidence of pest is observed in all crops and the families in villages of Seer Govardhanpur, Sarai Dongre, Lohta, Chitaipur and Nathupur, in general, estimate 25 per cent crop being affected by this constraint. The worst affected are paddy (30 per cent), mustard (70 per cent) and vegetables.

Animal-related

- The communities of the villages of Seer Govardhanpur, Sarai Dongre, Lohta, Chitaipur and Nathupur reported that the damage caused by blue bull (as pointed by two sample muhallas/hamlets) result in reduced cropping area with sizeable land left uncultivated.

- In Tikri, major agricultural constraint is damage due to blue bull.
- In Navampura, The crop damage by blue bull has also increased sharply since 1978.

Disease and Pollution-related

- Banjha (sterility) was reported in tomatoes, brinjal and pigeon pea crops by village groups of Tikri and Maraon (1 or 2 plants in 100).
- Tikri people reported that pollution is there. Air pollution—due to satellites in the sky. Land pollution—due to increased garbage resulting in cholera. Water pollution—Until 2-3 years back it was neutralised by the self-cleansing power of river Ganga. In recent years Ganga's magic has declined.

Infrastructure/Input-related

- In Navampura, vegetable cultivation is problematic due to lack of irrigation facilities.
- In village Navampura, some farmers are aware of the decreased fertility of land as a result of increased use of fertilisers. Some farmers are of the opinion that soil fertility has declined because of the irregular weather that changes with rapidity.
- For agriculture, summer season was not utilised to the full in the villages due to poor irrigation facility and electricity diverted for threshing of harvested wheat.
- In Chandpur, due to land acquisition and purchase by industrial estate, the production of vegetables in Chandpur has increased while that of crops has declined. One major reason being reduction in size of land holdings, which affects economic viability of crop production.
- In Tikri and Maraon, timely availability of water was perceived to be a major constraint.

Note: As described by women groups and men groups from selected villages of Varanasi

Source: Field Reports of PA Researchers from Varanasi

Coping Strategies

The local communities, in the urban and peri-urban areas of Varanasi, have adopted diverse coping strategies to hedge against low returns on agriculture, high risks of crop cultivation and uncertainty of wage employment. In the villages under study, some of the problems/constraints were most often overcome by personal efforts. These include;

- arrangement of generators to make up for electricity disruptions;
- increasing the amount of fertilisers and pesticides to increase the crop yield;
- construction of high boundary walls to ward off nilgai menace;
- changing cropping pattern for hedging against nilgai menace;
- changing cropping pattern for reducing the risk of disease and pest attacks, for e.g. reducing the cultivation of sugar cane and selected vegetables and growing wheat; and
- exploring alternative sources of livelihood in addition to agriculture.

2.4 Impact of Air Pollution at Varanasi

The women and men groups at Varanasi described the sources of local pollution more to brick kilns, domestic coal stoves, garbage and less to vehicular pollution. The women and men groups are aware of selected impacts of air pollution upon village communities, as shown in table 2.4. In general, the village communities are able to relate the direct impact of air pollution more with their health problems and less with their agricultural problems, unless it is an impact of pollution quite close to their field, for e.g. that caused by a brick kiln near-by. Scientifically, though there exist evidence of one to one relations between certain pollutants and certain kinds of disease, weeds, pest attacks adversely impacting on crop yield and growth, the village communities have different perspectives in this regard. The PA researchers have observed that the incidence of pest, insect and weed is noticed more in those villages, which are situated in the ozone pollutant zone, where dependence on farming is greater.

Table 2.4   Pollution—Sources and Impacts, Varanasi

Name of Village	Possible Impact Identified	Selected Observations and Present Measure/Action
Seer Govardhanpur, Sarai Dongre, Lohta, Chitaipur and Nathupur	• Insects and weeds are reported in paddy and wheat accounting for near total yield loss. • As per women groups, vehicular pollution is affecting crop production and health of human beings. • Damage to yield of wheat due to leaf injury, in the 5 villages, has so far been marginal (0.5-2.0 per cent) and the farmers do not consider it as a constraint. • Aphid pest attack in mustard is causing significant yield loss and high damage. • Impact of yellow spots in wheat on yield is marginal, so the farmers ignore it. • Mustard and paddy are the worst affected amongst the crops grown.	• The men groups did not relate impact of pollution on human health. • The women opined that pollution has increased since ten years though low on priority. • The local communities placed more importance on access to input (irrigation and seeds/ fertilisers) followed by pests-related problems and constraints. • The women and men are unable to address all the problems related to weeds. However, they apply medicines for pest and insect attack with advice from private shops. • Majority of farmers do not perceive pollution either as a constraint or a factor influencing it. They consider the current constraint in agricultural production as

| Chandpur, Navampura Kala | • In Chandpur, pest attacks in cultivation of vegetables have increased over the past 10-12 years.
• In Chandpur, though the brick kilns have closed, fruit production has dropped and soil is damaged.
• In Chandpur, the chemical effluent in the form of wastewater, from industrial estate and saree printing units have made 10 bighas of land barren. | resulting from defects in monsoon, inferior quality of soil/ seeds and lack of irrigation. Only 2 groups (total 16) in the 5 villages covered consider pollution as a constraint that leads to dust deposition on the crop due to vehicular movement. Another 5 groups think that the constraints, influenced by easterly wind (carrying primary pollutant from vehicular movements on grand trunk road) affect crop yields. The women groups perceive that fertiliser inputs influence the impact of pollution on crops.
• There is little or no awareness of the adverse effects of air pollution in the hamlets of Chandpur, possibly because there is no primary pollutant near-by. |

	• The villagers are aware of the adverse effects of emissions from the brick kiln (Chandpur village) on mango (blackening in colour, lessening of its sweetness etc.), on papaya (lessening of its sweetness) and on neem tree(making the leaves yellowish and weakening its twigs and small branches).	
Tikri, Maraon and Maheshpur	• Villagers reported loss in yield of fruit from the orchard (mango) due to smoke coming from brick kiln. Loss in mango and berry yields due to the deposition of smoke etc. from the brick kiln in the neighbouring villages.	• Vehicular pollution was not identified as a source of pollutant in villages due to less frequent vehicular traffic in absence of roads, open space and greenery in the villages.
	• Pollution from brick kiln was one identified source of pollution in Maraon village. One group said that due to smoke from brick kiln fruit yields (mango, guava) drop and the taste is bitter. The group also talked about pollution due to pesticide and fertiliser use in soil. These chemicals contaminate their food and weaken their health.	• Two men groups of Maraon, however, correlated increased incidence of pest attack, disease and weather changes due to pollution from the small factories in the town.
	• Farmers of Maraon reported that around 5-6 years back (when steam engine was still running), their vegetables usually cauliflower used to get damaged with 50 per cent loss (as it turned black due to the fumes). They also reported fall in yield of wheat due to the same reason.	• In Tikri, one group said that until 5 years back they were using ashes to prevent pest attack. But now it has become ineffective as the pest attack has taken virulent form. Hence they are forced to use pesticides, which are effective but expensive and increasingly require higher doses.
	• Tikri, did not report any increased pest attack (given the same seed variety in consecutive years) except for one ward where pest attack had increased (which had	

used different variety of seeds) over the years. The latter resulted in more pesticide consumption. The farmers thought that this was due to increased resistance to pests.

- Chlorosis was observed on cauliflower leaves in the southern fields of Maraon village.
- In Maraon village, there was consistent rise in pest attacks over 1990-95, when pest attack was virulent thus forcing some of the farmers to stop growing cauliflower (1994-96). But after 1996, such intensity fell and they started growing it again. Now, this year the pests have increased. The rise was ascribed to weather fluctuations.

- The villagers felt that the continuous exposure of women to smoke from coal stoves for cooking made them oblivious to any change in air quality in recent years, though they reported significant change in city's air quality.
- In Tikri village, one group observed that no diseases were prevalent earlier but since 1970s a variety of diseases have appeared, which have caused much damage to crops. Such trend was attributed to excess fertilisers and change in weather.
- It is important to note that though the farmers of Maraon raised their pesticide quantity, they always gave under-dose of it.
- Pollution due to garbage was identified as one of the areas of concern in the villages of Tikri, Maheshpur and Maraon. It was reported to be the causal factor for poor health of human and livestock. In Maheshpur and Maraon, it was also considered to be a major factor causing growth in pests.

Adityanagar, Karamanbir, Nuaon, Tarapur

- A women group from village Tarapur, Varanasi, stated that smoke from brick kiln adversely impacts on mango yield and leads to blackening of its outer surface. According to them, the yield of mango has reduced drastically over the last 10 years. Brick kilns, mainly surrounded by agricultural fields, lead to increase in temperature of the adjacent environment. They also reduced yield by 10 per cent to 20 per cent.
- The men group from Nuoan and Karmanbir villages were concerned about problems in farming activity undertaken in proximity to brick kilns such as plant diseases in 'Gerua rog' in which, the entire plant turns pale and gradually dies and affects yield. Mango fruit is adversely affected with fruits falling off before ripening and fall in both yield and taste. In case of lemon, the plant gradually dries up with reduction in fruit size and juice. There is rapid erosion of fertile top soil in the agricultural lands adjacent to brick kilns. The problem of erosion leads to creation of depressions in the field and this leads to problem of water logging especially in the rainy season. Such water logging decays crops. Even in other seasons agriculture becomes difficult on such lands and the farmers are forced to sell these

- One ward in Tikri reported purification of seed variety by self only to reduce pest attack and weeds.
- A women group of village Tarapur Varanasi denied any impact of vehicular emissions or the impact of smoke of brick kiln on human health. As per a Dalit woman, the dirty air of the city cannot reach here.
- The men group from Rajbhar hamlet, village Nuoan, stated that paddy and vegetables being rainy season crops are not affected significantly by smoke from brick kilns.
- The men group of Nuoan were also of the view that the loss of top soil and the loss of fertility can be recovered after thousand of years.

lands to brick kiln owners at prices
dictated by them.
- The women group from Patel,
  Rajbhar And Dalit hamlets of
  village Nuoan identified impacts
  such as the emission of hot air from
  brick kilns burns adjacent agriculture crop; the smoke coming out of
  the brick kilns causes blackening of
  the outer surface of fruits like
  mangoes with 50 per cent of the
  mangoes getting black; the yield
  has reduced considerably over the
  last 6 to 7 years; and the dust
  pollution of the vehicles also causes
  reduced yield, which is difficult to
  estimate.

Note: As described by women groups and men groups from selected villages of
            Varanasi
Source: Based on field reports on Varanasi by PA Researchers

2.5 Health and Social Issues at Varanasi

The women and men groups in the villages of Varanasi have listed a
number of diseases and illness, which they and their children suffer from.
There is increased incidence of cough, respiratory disease, stomach-related
disease, skin disease, eye infection, TB and malaria. Many children are also
victims of such diseases in addition to nutrition-related problems. As shown
in table 2.5, the village groups relate such diseases to a mix of factors such
as increased use of fertilisers and chemicals in agriculture causing stomach-
related disease/illness etc. and respiratory disease due to change of weather.
However, not many can perceive any impact of air pollution causing
serious impairment of health.

Table 2.5   Listing of Health Issues, Varanasi

Name of Village   **Type of Disease/Illness identified**

Seer Govardhanpur, Sarai Dongre, Lohta, Chitaipur and Nathupur
- The women perceive excessive use of fertilisers/pesticides, air pollution and change in weather causing new type of ailments.
- Some incidence of cough turning to T. B. and increased incidence of sickness.
- Women group in Sarai Dongre perceive the adverse health impact of vehicular movement (transporting sand from river Ganges).

Chandpur, Navampura Kala
- Increased incidence of cough, skin disease, stomach enlargement, and eye disease in children in last 15-20 years.
- Increase in incidence of skin disease, TB, gastric ulcer in adults, especially women.
- In Chandpur, rise in incidence of malaria due to mosquitoes.
- In Chandpur, the women group said that pneumonia had become endemic for the last 7 to 8 years and at least 100 children were affected last year, of which 10 children died.
- In Chandpur, health problems relate to cases of jaundice; incidence of gastritis (increase in incidence over the last 7 to 10 years supposedly due to contaminated food); frequent body ache and mild fever in elder members of the community; incidence of malaria (50 persons affected during last June/July) due to breeding of mosquitoes in stagnant water without proper drainage facilities; incidence of asthma (25 per cent of adult and old men and women are suffering for they think that the disease is incurable, medicines relieve temporarily); children affected by brain fever (taking 20-25 days for recovery and an expenditure of 6 to 7 thousands); increase in incidence of pimples all over body caused due to mosquito bites with 50 per cent of children suffering from this disease; incidence of children suffering from diarrhea for which hinga (Asafetida) mixed with milk is given for drink; children suffering from ear disease with pimple in the ear and pus flowing continuously out of ear; and nagging cold and cough affecting many persons.

Tikri, Maraon and Maheshpur

*   Groups from Tikri and Maraon villages pointed out the poor health of the villagers, especially the younger group, because of increased consumption of fertiliser-based crops and poor food intake.
*   Tikri village highlighted pre-mature graying of hair, no reason ascribed. Some villagers of Tikri village thought that TB and asthma were hereditary diseases.
*   The villagers spoke of poor health amongst the younger generations in both Tikri and Maraon, due to consumption of crops based on chemical fertilisers.
*   A moderate increase in respiratory problems was reported from Tikri and Maraon. As explained by the villagers, better accessibility of doctors as compared to earlier days, increased consciousness and knowledge of the ailments previously unknown.
*   In Morya helmet of Tikri village, one health problem is related to stomach trouble due to consumption and handling of grains to which insecticides and pesticides have been applied.
*   In Morya helmet of Tikri village, some villagers are getting sores on different parts of the body, such as hand, head, face, stomach, legs etc. The disease starts with patches of swelling, increasing gradually in size and getting converted to open sores. It usually takes six months or more for healing. The treatment for the sore is expensive costing a minimum of Rs.500/- per month per person. Around 5 to 6 persons in the village have been affected by such swelling, reasons for which were attributed to atmospheric temperature and changes in weather.
*   In Morya helmet, the participants thought that they were getting weaker due to consumption of food grown with the help of fertilisers and pesticides, which lacked nutrients and could not provide strength. Consumption of such grains by livestock also led to weakness in livestock; the milk yield of livestock got adversely affected and the quality of milk had low level of nutrients.

Adityanagar, Karamanbir, Nuaon, Tarapur

*   In village Nuoan, apart from common diseases like cholera, small pox, malaria etc. people have also complained about a recent problem of eye irritation especially during morning hours. They think that it may be due to long exposure of eyes to the atmosphere in which smoke as pollutant is discharged.
*   In village Adityanagar, some of the health problems are TB and asthma, diarrhea, cholera and malaria in children as well as in adults, pneumonia in children, eye problems in women and children, running nose in children.
*   In village Karamanbir, some of the health problems of women and men having high incidence malaria, TB, asthma, arthritis, stomach disorder and gas formation, diarrhea and eye problems.

- In village Tarapur, there is high incidence of malaria and asthma in men; stomach disorder, gas formation and eye problems in women; and malaria, pneumonia, chronic cough, eye problems, diarrhea and vomiting in children.

Note: As described by women groups and men groups from selected villages of Varanasi

Source: Based on field reports on Varanasi by PA Researchers

"Our" Social Issues

The women groups, in general, in the villages covered during field research, lack empowerment especially regarding access and control over funds and mobility for jobs. Girl children are discriminated in terms of schooling and social investment. The women work hard and have little time for recreation. The personal standards of hygiene and sanitation of some women groups are low and have little knowledge or empowerment to improve their conditions of living and quality of life. Many of them have a lowly state of existence where their husbands get drunk and often resort to wife beating. Their workload is often too heavy which affects their general state of health. Many women are vocal and sarcastic with very low threshold of tolerance and easily provocative. Their children are malnourished and under-nourished and without access to funds from the male family members the women find it extremely difficult to cope with the reality.

- The emphasis on education is little in these villages except amongst the upper caste. Education for girls is generally discontinued after the middle school so as to train them in household chores. In the caste hierarchy, children from the lower castes work in the field after school. Many girl children are forced to discontinue their schooling in order to look after their siblings.
- Some women groups including those from Chandpur village described their miserable state of existence. Though many men worked in the industrial estate and neighbouring town and earned enough to maintain family, few cared to contribute enough to the family.

- Drinking habit in the villages is rampant, where most men take alcohol including adolescent boys.
- Abuse and wife beating are very common in the villages where most men take little interest in either household activities or rearing up of children.

2.6 Agriculture Support System, Varanasi

Institutions identified by the farmers as part of their support system are

- Sarpanch, panchayat/village head;
- Panchayat Samiti;
- Block Development Office;
- Local agriculture departments and its staff;
- Mandi market actors (the farmers sell their produce);
- Local private traders from whom they buy agricultural inputs, (e.g. fertilisers, seed, pesticides, cash, latest information);
- Cooperatives/banks;
- Local community-based organizations; and
- Infrastructure departments of the government (electricity, water, irrigation, veterinary services, forestry, education, health etc.).

Varanasi—How Effective is the Support System for Farmers?

The villages did not have an effective government or community-based organization to guide and advise the community at the time of crisis. No support was available to them from outside and the villagers made their own arrangement to cope with agricultural problems/constraints such as nilgai menace, problem of pests and weeds, low quality seeds, increasing cost of fertilisers, erratic electric supply, natural calamities etc. The local community members generally reported their problems to the village head or accepted the state of affairs.

About the organizational support system, the farmers' service centre (Kisan Sewa Kendra), an initiative of the government, is expected to provide a forum for farmers to represent occupational issues to the administration on a regular basis for necessary support and action. Initiated in 1994 and held every week on Thursday, it serves as a post office to transmit issues raised by the farmers but not much action has happened in practice. However, in another Venn diagram drawn by farmers' group such Kisan Sewa Kendra was shown as an important institution. It appears that perspectives on Kisan Sewa Kendra vary across hamlets and socio-economic groups. This is because the Kendra has differential response to the issues raised by farmers and some, especially the small and marginal farmers do not find it of much relevance.

A Venn diagram drawn by one farmer's group in village Maraon shows their relationship with the various identified institutions—block development office, seed shop, fertiliser and pesticide shop, pump set owner, tractor owner, gram sewak, wholesale market, village lender and village chief. The input traders hold very important place as they act as sources of information. They are in direct touch with the government agricultural farm. In case of fall in yield the farmer consults them who in turn gives advise. Farmers trust them because of long time relationship and also the convenience in having this source of information. Farmers also depend a lot on opinions and advice of fellow farmers, friends and relatives.

The village money lender's role in village Maraon has declined over time due to diversification of household livelihood strategies. Any loss in agriculture is met by income from other sources. The villagers of Maraon went to the District Magistrate's office, early last year for getting tube wells installed in the village but no action has happened at the field level. Almost 3 years back, menace by stray cows led the cultivators to appeal to their owners in adjacent Police colony, which proved effective for the cows stopped coming. A Venn diagram drawn by one farmer's group in village Sarai Dongri shows that villagers attach high weights to private shops since it supports them with seeds, fertilisers and pesticides. Irrigation and electricity departments are the next highest in priority followed by cooperatives. In the Venn diagram, the role of the block office is perceived as negligible and the Kisan Sewa Kendra finds no mention in such diagram. In the absence of effective organizational support system, the farmers look for advice from the private traders in fertilisers/pesticides when faced with pest, disease and problems in crop yield.

Often with improper guidance by the traders, difficult situations arise for the farmers.

- Excess/low dosage of fertiliser/pesticide getting applied;
- Inappropriate fertiliser/pesticide getting applied; and
- Adulterated fertiliser/pesticide getting applied.

Such situation seems to cause toxicity of plant and animal life.

A government agriculture farm near the villages is involved in providing fertilisers to the farmers. However, only the powerful men of the village benefit more through higher procurement. The farm has also facility for testing soil, yet the officials of the farm are not cooperative and hence, the farmers rarely approach them.

Though farmers are members of cooperative society, its services are not available to them due to two factors—(a) most of the farmers are defaulters and (b) farm input supply is not in time, hence the question of borrowing does not arise.

The nearby bank branches are also not supportive and deny loans to farmers because of high rate of default of villagers on bank loans. The local block development office organizes farmers' meetings but the farmers are not aware and/or motivated to join such meetings.

In another Venn diagram drawn by men group from Adityanagar, institutions like Kashi Gramin Bank, primary school and Panchayat Bhawan were drawn closer and equidistant from the farmers, whereas, institutions like Police Station and Panchayat Samiti were drawn at a greater distance from the farmers. The latter indicated the dissatisfaction of the villagers with the two institutions. The reasons for expressing dissatisfaction with the local Police Station was that the Police officials were corrupt and demanded money at every opportunity. They also misbehaved with general public. While, the Panchayat Samiti was not fulfilling its duties/ responsibilities fully. The village Panchayat received only 60 per cent of the total grant from the government. Most of the villages in the jurisdiction of Panchayat Samiti were deprived of the various schemes undertaken by it. Further the Samiti had failed miserably to perform its welfare activities in the areas of providing medicines in wells, conducting immunisation programme, family planning and other related programmes.

As per perspective of women group from Adityanagar the administration was also responsible for acute shortage of water and insufficient electric supply. Corruption at public places was a major problem due to which poor people were deprived of the facilities offered by the government. The decisions of village Pradhan of Adityanagar favoured higher and middle class people in the area rather than poor and deprived groups.

In Nuaon and other villages, farmers complained about poor seed quality, which they received from Government officials. Since poor varieties of seed were sold to the farmers with labels of high yielding variety, the farmers found it difficult to assess the quality of seeds. The farmers were able to gauge such farce only after harvesting the crop and experiencing drastic reduction in crop yield. The farmers of village Nuaon were of the opinion that the Government can solve the problem by punishing authority involved in such affair. Fertilisers were often not available in the block office, which forced farmers to purchase fertilisers from local markets at high rates.

Some common organizational weaknesses and hurdles at Varanasi as mentioned by different farmers' groups were

- inefficiency and bankruptcy of cooperatives and the Kisan Sewa bank;
- corruption of government officials at the implementation level; and
- lack of communication between villagers and officials.

2.7 Impact of Urbanisation and Industrialisation on Quality of Life at Varanasi

In many hamlets, majority of people felt that the process of local industrialisation has helped them by offering more employment, a larger market and higher prices for their products such as milk and food grains. The women groups were particular in pointing out the costs of urbanisation and industrialisation as indicated below in table 2.6. The major costs being growing market-driven commercial crops; high doses of external inputs and high accompanying risks; high pest problems, food insecurity and problem of livelihood with declining role of agriculture. Migration of men created problems for women while there was pollution due to industrial waste causing fall in sanitation standards and social standards with high crime rate.

Table 2.6   Impact of Urbanisation and Industrialisation, Varanasi

Names of Villages	Benefits	Costs
Seer Govardhanpur, Sarai Dongre, Lohta, Chitaipur and Nathupur	• None perceived.	• (Women group) Change in cropping pattern is market driven and depends on external inputs and large invest ments. In the event of crop loss, this leads to a net loss due to considerable gap between income and expenditure. This also subjects the farmers to enormous difficulties in realising of proportionate return from agriculture. The women headed households and marginal farmers are the worst affected in such circum-stances. Earlier, the households grew a number of food crops, which spread the risk. But with reduction in the number of food crops grown, the risk is high. Now growing paddy, though preferred, is risky since it is most affected by pests. Land acquisition by government contributed to displacement of occupation from agriculture to un-skilled labour. It also contributed to social

		pressure on women through migration. The current trend is towards fall in share of agricultural land due to land acquisition and construction activities.
Chandpur, Navampura Kala	• (Men group, Chandpur village)—Employment gets generated; children also get employment; employment saves family from starvation; possibility of income increases by selling vegetables, tea and others during lunch time to the factories. • (Women group, Chandpur village) Land becomes costly, income from land is enough to raise standards of living.	• (Men Group, Chandpur village)—Crime rate increases. Outsiders engaged in crime such as robbery, picking pockets etc. (Women group, Chandpur village) Villages become dirty due to excess garbage thrown from industries; crime rate increases; natural air gets contaminated due to crowd, dirty water and excess garbage; with breeding of mosquitoes the incidence of malaria increases; dirty water also leads to diarrhea;
Tikri, Maraon and Maheshpur	(Men group, Tikri village)— • Industries create a very good market where local villagers get high price for local products, such as milk and vegetables. • The land becomes costly, hence higher returns accrue from sale of land. • Local persons get more employment at lower levels as labourer. (Men group from Morya hamlet, Tikri village) • Land becomes costly for ordinary persons.	• None perceived.

- Those areas near factories
  get crowded.
- Milk and vegetables get
  sold at higher prices.
  (Men group, Maraon Village)
- One benefit perceived by
  the group is ready market
  for their vegetables.
- Another benefit is that of
  providing employment.

Note: As described by women groups and men groups from selected villages of Varanasi
Source: Based on field reports on Varanasi by PA Researchers

2.8 "The Do-able's" at Varanasi

"The do-able's" suggested by the women and men groups mainly relate to overcoming the infrastructure/input constraints that adversely impact their agriculture. Such constraints have been discussed above. The women and men groups are of the opinion that their priority problems related to infrastructure and input can be resolved by means of action taken by the government, local communities and non-governmental organisations. Some of the "do-ables" suggested by the women and men groups from the villages at Varanasi are as follows.

For improvement of agriculture, the Government can adopt the following measures.

- Can ensure adequate supply of electricity for irrigating crops.
- Can provide fertilisers and medicines at subsidized rate.
- Can take disciplinary actions against manufactures of duplicate medicines.
- Can make loans available at concessional rates.

The other kinds of "do-able's" suggested are by the PA researchers on the basis of their interactions with the community and their personal observations.

Selected "Do-able's"

- There is need to examine the conflicting issues between the law of protection of wild life especially the engendered species of nilgai and protection of agriculture crop as a livelihood. It is important to work out a mutually beneficial system.

- There is the need to formulate policies that promote sustainable agriculture with agro-economic processes that reduce the impact of air pollution on crop production. The benefits of green manure can be disseminated for improving the soil quality and fertility of land. It is important to build community awareness on the issue and involve them in policy making so as to incorporate their knowledge and views.

- Community awareness building on impacts of air pollution and ways of mitigating them are essential since at present the community does not perceive the impact of air pollution and the attention is focussed more on input supply.

- With dwindling size of land, cooperative farming or consolidated farming practices can be adopted so as to increase the productivity of farming.

- There is clear need for policy dialogue amongst the policy makers, people and other actors on the aspects of food security vis-a-vis food production. In this context it is important to recognise the implications of such a policy for the poor and deprived households who are the worst affected in the process.

- Agricultural extension regarding better farming practices including judicious use of pesticides and fertilisers should be provided. The concept of seed dressing should be introduced whereby seed should be checked for quality and pesticide content before releasing it for sale.

- Consumer courts should also be strengthened so that local community is empowered to ask for compensation for the damage done by air pollution from polluting units.

- For better utilisation of land, agro-forestry can be introduced in those lands left fallow due to nilgai menace or some other agricultural constraint. Plantation of trees can also help in mitigating pollution.

- Irrigation facilities can be improved by constructing minor irrigation channels.
- Lessons in women's empowerment can be gradually undertaken by development agencies, whether government or non-government.

Annex 'A' Profile of Villages at Varanasi

Name of Village   —   Adityanagar
Name of Block   —   Kashi Vidyapeeth
Name of District   —   Varanasi

Socio-Economic Profile Village Adityanagar is one of the scientific sites of Benaras Hindu University and is highly polluted area due to vehicular pollution. It is a roadside village located on Pachkosi road and has 5 hamlets. It is a part of Karaundhi Gram Sabha, which falls under Kashi Vidyapeeth block. It has a population of 8000 with 900 households. Most of the villagers residing in village Adityanagar belong to Kunbi caste and are economically weak. It is an urban area and has modern infrastructure/ facilities on the roadside such as multi-storey buildings, medical practitioner. Studio, hotel, grocery and cloth shop. About one quarter of the village (north of Pachkoshi road) is under municipal area and three quarter of the village (south of Pachkoshi road) is non- municipal area.

Agriculture and Livelihood The main sources of livelihood for men are wage labour (nearly 180 households are engaged in it), hawkers, embroidery work and shop keeping. For women of the village it is garland making from beads and beedi rolling. The villagers grow vegetables in their home garden. Men are engaged in construction activities as wage labour. Around 2 per cent do share-cropping in near-by villages. Many women perceive the importance of agriculture because their parental residence is in agriculture-oriented villages.

Issues Related to Air Pollution The women have practical problems tackling pollution sources such as domestic garbage and water flowing out of flooded sewage. They are not really aware of the impacts of vehicular pollution.

The village groups suggested a major role of the government in addressing their agricultural constraints and general problems of the village.

Name of Village   —   Chandpur
Name of Block   —   Kashi Vidyapeeth
Name of District   —   Varanasi

Socio-Economic Profile Chandpur is a large sized village situated near Lahartara industrial estate. The village is a scientific site of Banaras Hindu University and has relatively high levels of pollutants such as nitrogen dioxide, sulphur dioxide and ozone. The village has a total population of 5682 with 255 households. 'Patel' is the dominant caste in the village and the main occupation is agriculture and wage labour. The village has two private junior high schools, a primary government school, a bank, a post office, a Kisan Kendra, a government agriculture farm and a Block office.

Agriculture and Livelihood Local livelihoods consist of farming and livestock raising, services, wage labour including agricultural labour and other activities such as saree printing, processing "Rudraksh" (a holy bead), preparing 'bidi' etc. The village has 40 bigha's of agricultural land and grows wheat, paddy, jowar, corn etc. mainly for self- consumption. The vegetables grown are sold in the market. Around 30 per cent of total income is derived from agriculture.

Issues in Agriculture There is rise in incidence of pests and diseases in crops and vegetables; fall in crop yield with use of chemical fertilisers; 10 bighas of land in the village has been made barren due to flow of chemical water from industrial unit; activity of brick kiln has made land unusable; brick kilns have also affected horticulture and reduced yields of mango, jackfruit and guava which now have high rate of termite attack; at least 5 bighas of horticultural land have been converted for vegetable cultivation; water bodies in the village are adversely affected due to air pollution and they do not grow 'singara' (a green colour triangle-shaped fruit in water plant) any longer. Earlier such fruits were grown and sold on a commercial basis.

Issues Related to Air Pollution Incidence of children falling sick have increased; the children tend to have prolonged cough; they also have problems with their eyes; there is incidence of TB, which was not there earlier.

Some suggestions by village groups are that production of green manure needs to be promoted in order to raise the productivity of land. More trees can be planted rather than felled thus reducing pollution levels. Cooperative farming can be introduced in order to overcome the constraint of uneconomic land size. The farmers can be made aware of technical knowledge through agriculture extension. Free bore wells can be provided to small and marginal farmers. New technologies for irrigation can be promoted. The Block office or the banks can find ways to resolve the problem of credit.

Name of Village	—	Chitaipur
Name of Block	—	Kashi Vidyapeeth
Name of District	—	Varanasi

Socio-Economic Profile Village Chitaipur has Nitrogen Dioxide and Sulphur Dioxide as primary pollutants in the village.

Agriculture and Livelihood Village Chitaipur has clay soil. Agriculture and agriculture labour contributes 90 per cent as a source of livelihood with livestock 1 per cent, services 3 per cent and business 6 per cent. In village Chitaipur, the Kharif season has paddy growing on 75 per cent of the land, blackgram, greengram, brinjal and ladies finger growing on 15 per cent and maize on 10 per cent of the land. During the Rabi season, wheat grows on 75 per cent of the land, vegetables like brinjal, radish, potato and tomato grows on 20 per cent of the land and the rest of the 5 per cent of the land has gram, pea and mustard growing on it.

Name of Village	—	Karamanbir
Name of Block	—	Kashi Vidyapeeth
Name of District	—	Varanasi

Socio-Economic Profile Village Karamanbir is located 3 kilometers west of Benaras Hindu University. Villages' Nuaon, Brindaban and Susuwahi on the north, south and west sides surround it, respectively. A social cum resource map drawn by a village group of Karamanbir is given in PRA chart 3.1. There are three hamlets in the village, Khanpokhari, an urbanised portion on the road side, Bichalapura hamlet joins Khanpokhari and Karamanbir Brinda, located in the interior part of the village is a rural area. It is one of the scientific sites of Benaras Hindu University and is rated as highly polluted area due to vehicular pollution.

Agriculture and Livelihood In village Karamanbir, about 50 per cent of total land is agricultural land, out of which 75 per cent is low land and 25 per cent is high. The vegetables are grown on the highland round the year and paddy and wheat are grown on the low land. The farmers are practising mixed cropping, especially of winter crops such as mustard, wheat and peas.

Issues in Agriculture Selected issues raised by women were nilgai menace, acute shortage of electricity supply for irrigation, low lying land for cultivation, increasing amount of fertiliser inputs and coping with many crop diseases, insects and weeds and natural calamities like excessive rains and hailstorm.

Issues Related to Air Pollution Village Karamanbir being adjacent to village Nuaon faces the problem of air pollution from emissions through brick kilns, though its impact is less severe. Most villagers do not consider air pollution as a big issue and are not aware of the impact of vehicular pollution. Some have raised the issue of black spots in mangoes due to air pollution.

According to village groups the government can play an effective role in solving their agriculture problems and constraints by providing agriculture inputs at subsidized prices, provide loans, ensure regular supply of electricity and tackle the nilgai menace.

Name of Village	—	Lohta
Name of Block	—	Kashi Vidyapeeth
Name of District	—	Varanasi

Socio-Economic Profile Village Lohta is a Muslim village and it is the centre of Banarasi saree weaving. This village has Ozone as air pollutant as per the scientific data.

Agriculture and Livelihood Village residents are mainly engaged in saree weaving. Agriculture crops are grown during kharif and rabi season. During kharif season crops grown are paddy and brinjal. Rabi crops grown are wheat and mustard and vegetables like tomato, brinjal, radish, potato and chilies.

Name of Village	—	Maheshpur
Name of Block	—	Kashi Vidyapeeth
Name of District	—	Varanasi

Socio-Economic Profile In village Maheshpur, there are 500 households with a total population of 5000. It is one of the scientific sites of Benaras Hindu University where levels of Nitrogen Dioxide and Sulphur Dioxide are high. Maheshpur is a roadside village and is large-sized. The Grand Truck (G.T.) road passes through the village. Lahartala Industrial Estate is less than 2 kilometers south west of the village. In Shivdaspur side of the village no household is engaged in agriculture. In Chorwa bari part of the village only 25 households in Yadav Basti own some land and do farming.

Agriculture and Livelihood One dominant caste is that of 'Patels' and main occupational sources are that of labourers and business. The village community cultivates vegetables on a commercial basis and grows cereals for personal use. Total cultivable area in the village is 4-5 acres. In the rainy season crops like paddy, maize, pigeon pea, bajra, horse beans, moong, ninwa, bitter gourd, radish and spinach are produced. Spinach is the main commercial crop and bajra is the fodder crop. In the winter season, brinjal, cauliflower, spinach, radish, tomato, wheat, pea and potato are sown. The wheat is grown for personal consumption.

Issues in Agriculture Some issues in agriculture raised by village groups were problem of irrigation; availability of fertilisers for commercial crops; adverse climatic conditions like excess rain, unseasonal showers, fog and frost; crop damage by porcupines, nilgai, stray cows and rats; crop diseases, weeds and pests; problem of land ceiling act and forced sale of land; and shortage of manpower.

Issues Related to Air Pollution The villagers reported mango trees getting affected by smoke from nearby chimneys of iron moulding factories and brick kilns in their village. Mysterious blackening of paddy ear-rings were observed last year in the season of kuar, which resulted in 25 per cent yield loss.

Name of Village	—	Maraon
Name of Block	—	Kashi Vidyapeeth
Name of District	—	Varanasi

Socio-Economic Profile Village Maraon is a scientific site of Banaras Hindu University, which is situated approximately 4.8 kilometers north east to the

Benaras Hindu University (BHU). It shares its border with Nakain village in its south, Dafarpur village in its east, Faridpur and Hasanpur villages in south east, Hanharpur village in its north and Jalalipatti colony in its west. There are 17 brick kilns in the neighbouring village Bhulanpur. A railway line passes through Maraon village. Diesel Locomotive Works colony is 1.8 kilometers northeast from the village. Its total area is 300 acres. It is one of the principal scientific sites of BHU with high levels of sulphur dioxide and nitrogen oxide and concentration of ozone. Village Maraon has a total population of 2500 with 250 households. It has 7 wards. The dominant caste is that of 'Dalit' and it has been declared as Ambedkar Gaon (village).

Agriculture and Livelihood Sources of main livelihood in village Maraon are agriculture and wage labour. Its total cultivable area is 260 acres. About 75 per cent of the population are engaged in agriculture. In summer 40 per cent of land is kept fallow and on 60 per cent sugarcane and vegetables are grown. In the rainy season, paddy is grown on nearly 30 per cent of land, the rest of the land is devoted to growing vegetables and jowar and bajra and only 5 per cent is kept fallow. In winter, wheat and mustard is grown on 70 per cent of the land and mainly vegetables are grown in the remaining land. Milch animals like buffaloes are an important source of livelihood. Rearing of pigs, goats and hens is also important. The village specialises in cauliflower production. 40 per cent of women supplement income by making necklaces of artificial beads and Rudraksha (a fruit which hardens on ripening and used for making beads) and rolling bidis.

Issues in Agriculture The main problems of agriculture as faced by the farmers are pest attacks, which has increased over time, damage to crops from rain, hailstorms, frosts and fog; crop diseases and shortage of irrigation facilities.

Issues Related to Air Pollution Some water and air polluting factories were identified by the village groups such as brick kilns, tyre factories, bead manufacturing factories, machine tool factories, cement factories, railways, saree printing factories and indigo factories. Such pollution leads to deterioration of air and water quality and change climate. They lower the mango yield and size. It also leads to yellowing of wheat leaves and smaller wheat ear-rings. There are black deposits on cauli flowers and 50 per cent loss of crop. Due to air pollution, crops get weak, diseased and pest-prone. Fumes from nearby brick kilns adversely affect health, livestock and crops.

The village groups suggested a major role of the government for addressing their agricultural constraints and a role for local municipality for resolving problems like drinking water.

Name of Village   —   Nathupur
Name of Block   —   Kashi Vidyapeeth
Name of District   —   Varanasi

Agriculture and Livelihood In village Nathupur, agriculture is a minor source of livelihood, wage labour being the major one. The village grows kharif crops like paddy, pigeon pea, green gram, black gram, millet and pearl millet and vegetable like brinjal. Rabi crops grown are wheat, mustard, gram, pea, linseed and lentils and a whole range of vegetables like radish, tomato, potato, cauliflower, chilies, coriander, garlic and carrot.

Name of Village   —   Navampur Kalan
Name of Block   —   Kashi Vidyapeeth
Name of District   —   Varanasi

Socio-Economic Profile Navampur Kalan is a roadside village with the Grand Trunk road near to the village. It has a total population of 5000 with the number of households being 410. The dominant caste is that of 'Yadav', engaged in milk business. The village has a Panchayat office, post office, veterinary dispensary, high school, bank, seed and a fertiliser cooperative.

Agriculture and Livelihood Agriculture and livelihood are mainly livestock raising and some farming, services, trading, wage labour including agricultural labour and other activities such as making of cow dung cakes and shop keeping. As compared to past, agricultural land has increased and mainly wheat is grown on them. Such cultivable land is far from the farmers' homes. A few households produce vegetables in their home garden for self consumption.

Issues in Agriculture The issues in agriculture are irregularity of rainfall; lack of irrigation facilities in time; problem of disease and pest in crops; problem of applying large doses of fertilisers; inability to get fertilisers and seeds at reasonable rates; lack of credit and training in agriculture; erratic supply of electricity; no contact with officials from agriculture department;

problem of 'nilgai' (bluebull) menace; problem of flood and termite attack on land; high irrigation charges and high interest on credit.

Issues Related to Air Pollution Increased incidence of cough, TB and eye disease are in both children and adults since 10-15 years. Gastric problem exists since 8 to 10 years and incidence of skin disease in both children and adults, has increased since 5 to 6 years.

Name of Village   —   Nuaon
Name of Block   —   Kashi Vidyapeeth
Name of District   —   Varanasi

Socio-Economic Profile Village Nuaon is located 9 kilometers south west of Benaras Hindu University (BHU). It is surrounded by villages such as Brindaban on the east, Dafi in the west and Susuwahi in the north. Nuaon is spread over 415 acres of land with a population of 2200 and has 350 households. It has four caste-based hamlets. Village Nuaon is one of the scientific sites of BHU for studying the impacts of air pollution on selected plants. Most of the habitation area of the village and some of the agricultural fields are located on the Mugalsarai-Mohansarai bypass road. Due to the road, the urbanisation process has become very rapid. Only 30 per cent of the village area is residential while 70 per cent is agricultural land including roads and orchards.

Agriculture and Livelihood Majority of the population have experience in agriculture either as landowners or as agricultural labourers. Many households own milch animals and are engaged in producing milk. During rainy season the crops grown are paddy in 55 per cent of agricultural fields, animal fodder in 10 per cent, pulses in 7 per cent, sugarcane in 7 per cent, maize in 4 per cent, jowar and bajra in 2 per cent and vegetables in 15 per cent. In winter season, vegetables are grown in 35 per cent of the fields while wheat in 50 per cent and chana and peas in 15 per cent. In summer, 8 per cent of land is kept fallow and in the rest of the land sugarcane, vegetables and animal fodder are grown.

Issues in Agriculture The major agricultural constraints and problems are those of reduction in yield due to fog, hailstorm and other natural calamities, nilgai and rodent menace, decreasing fertility of land, crop diseases and weeds and insects.

Issues Related to Air Pollution The villagers are exposed to vehicular pollution, brick kiln smoke and heat and indoor pollution from traditional stove. They are concerned about ill-effects of brick kilns on agricultural activities like plant diseases mainly in wheat called 'gerua rog', which reduces yield and dries up the plant. There is also the impact on fertile top soil of the agricultural fields which get eroded due to brick kilns. The impact of brick kiln is negligible in the rainy season.

The village groups suggested a major role of the government in addressing their agricultural constraints and general problems of the village.

Name of Village	—	Sarai Dongre
Name of Block	—	Kashi Vidyapeeth
Name of District	—	Varanasi

Socio-Economic Profile In village Sarai Dongre, one air pollutant is ozone as specified by scientific studies.

Agriculture and Livelihood In village Sarai Dongre, agriculture and livestock constitute nearly 55 per cent as source of livelihood, wages make up for 38 per cent, services for 5 per cent and rural artisan and trade for 2 per cent of livelihood. Income-wise, agriculture contributes 25 per cent of total village income, followed by services, livestock and wages making for 40 per cent, 25 per cent and 10 per cent respectively. The kharif crops grown in the village are paddy, pigeon pea, green gram, black gram, millet and pearl millet. During Rabi, the crops grown are wheat, gram, pea, lentil and mustard and vegetables like potato, onion, lentil, garlic, chilies and coriander. In summer, the crops are sugar cane and melon.

Name of Village	—	Seer Govardhanpur
Name of Block	—	Kashi Vidyapeeth
Name of District	—	Varanasi

Socio-Economic Profile In village Seer Govardhanpur the dominant caste is 'Yadav'. It is a roadside village. It is one of the scientific sites of BHU where primary pollutants are Nitrogen Dioxide and Sulphur Dioxide.

Agriculture and Livelihood In village Seer Govardhanpur agriculture and livestock rearing as sources of livelihood make for 60 per cent, wages contribute 10 per cent, services makes for 20 per cent and business for 10 per cent. The cropping pattern of major crops in the village in summer are nenua, ladies finger and brinjal in 10 per cent of the land with 90 per cent left as fallow land. In Kharif season, the crops grown are millet and pearl millet in 60 per cent of the land, maize, green gram and pigeon pea in 30 per cent of the land and 10 per cent of the land remains fallow. In Rabi season, wheat is grown in 80 per cent of the land while 20 per cent of the land is used for growing mustard, gram, pea and lintel.

Name of Village   —   Tarapur
Name of Block   —   Kashi Vidyapeeth
Name of District   —   Varanasi

Socio-Economic Profile Village Tarapur is about 8 kilometers south of Benaras Hindu University (BHU). It is surrounded by Saraidangari on the north, Muradeo in the east and the river Ganges in the south. It is spread over an area of 535 acres with 170 households and 2400 population. It has 4 caste-based hamlets. The village has high level of ozone and vehicular pollution as per scientific data from BHU. Heavy traffic bypass is around 4 kilometers away from the village. It has many castes residing in the village. It has a primary school.

Agriculture and Livelihood Majority of the residents of village Tarapur are well acquainted with agriculture — either tilling their own land or working as agriculture labour. Dalits mostly work as wage labour in sand mining, in agriculture and construction activities. Land distribution in the village is very uneven with Bhumiars with 50 households owning 400 acres of land and Dalits with 40 households owning about 5 acres of land. During rainy seasons, kharif crops like paddy, maize, arhar, bajra and vegetables are grown. During rabi season, crops like wheat, chana, pea, mustard and potatoes are grown.

Issues in Agriculture The problems of agriculture are nilgai menace, insufficient electric supply affecting irrigation, increased insects and diseases in plants, natural calamities and shortage of labour.

Issues Related to Air Pollution The women groups perceived the impact of smoke from brick kilns, how it affects mango yields by 10 per cent to 20 per cent and blackens its surface. The women groups denied any impact of vehicular or other types of air pollution.

The village groups suggested a major role for the government in addressing their agricultural constraints and general problems of the village.

Name of Village	—	Tikri
Name of Block	—	Kashi Vidyapeeth
Name of District	—	Varanasi

Socio-Economic Profile Village Tikri is 2.04 kilometers south west of Banaras Hindu University (BHU). River Ganges is 2 kilometers east of the village. It is one of the scientific sites of BHU with air pollution such as ozone, nitrogen dioxide and sulphur dioxide. It is a large-sized village with a total population of 4,500 and total households numbering 450. It has a total number of 13 caste-based wards and a total area of 1500 acres. Nearly 40 per cent of population is literate in the village. The village has been recently declared as Ambedkar Gaon, since the scheduled caste population in the village is more than 50 per cent.The village has one government-run senior secondary school, an agriculture cooperative society, one branch of the Union Bank of India and one health centre (which is non-operational). The village has no industries. Its neighbouring villages have brick kilns with Narottampur having 2, village Kuruwa having 1, village Akhri having 3 and village Nuar having 2.

Agriculture and Livelihood In village Tikri, at least 70 per cent of population is engaged in agriculture and 50 per cent of agricultural land is irrigated. Selling milk is also a prime source of livelihood. The total cultivable area is 800 acres. It has two types of soil -alluvial and domat and hence, the cropping pattern is also soil-specific. There are three types of cultivators in the village, the landowners doing their own cultivation, the sharecroppers and those renting land on an annual basis. In winter season, double cropping is generally practised with vegetables like potatoes in early winter for commercial production and wheat in late winter for self-consumption.

Issues in Agriculture Some issues in agriculture for village Tikri are water shortage for cultivation, hailstorm, unseasonal rains, frost and fog, nilgai and rat menace, plant diseases and pests and low quality seeds.

Issues Related to Air Pollution There is no visible effect of vehicular pollution perceived by the villagers. Smoke and heat from brick kilns of neighbouring village burn the mango flowers and affects mango yield. Dust from brick kilns falls on mangoes resulting in black spots and its small size.

3 Farmers' Perspectives from Faridabad: A Green Revolution Belt in a Developed State

Introduction

Chapter 3 is divided into nine sections. Section 3.1 relates to perspectives of agricultural communities from Faridabad on the role of agriculture. Section 3.2 is on farming practices and seasonal activity calendars of such agricultural communities. Section 3.3 relates to their agricultural problems and constraints and their priority issues. Section 3.4 describes their perspectives on impact of air pollution on agriculture. Section 3.5 contains health and other social issues of such communities, while section 3.6 outlines effectiveness of their support system. Section 3.7 is a community-based assessment of urbanisation and industrialisation. Section 3.8 provides a description of the "do-ables" at Faridabad and Annex 'B' contains profiles of agricultural communities in urban and peri-urban areas of Faridabad. Selected PRA charts on Faridabad are given in Appendix 1 and the locational map of villages in Appendix 2.

3.1 Faridabad—Background

Faridabad district, in the state of Haryana, India was a part of Gurgaon district in Haryana until 1979, when it was established as a separate district. It shares its western boundary with Gurgaon, Delhi in the north and Uttar Pradesh in its east, north-east and south-east. Faridabad district has five blocks—Faridabad, Ballavgarh, Palwal, Hodal and Hathin with 62 villages, 81 villages, 122 villages, 70 villages and 79 villages respectively. As per 1991 Census Report of Government of India, of the total area of Faridabad district, rural area comprises of 1891 square kilometers and urban area of 189 square kilometers. The total cultivable area is 1.68 lakh hectare, of which, irrigated area is 1.39 lakh hectare. The total food grain produced is 399200 tonnes.

Faridabad is one district, which is a part of the success story of 'green revolution' in agriculture in the late sixties. With high yielding variety of seed, high doses of fertilisers and pesticides, better irrigation facilities and flood control measures, agricultural production reached heights supported by animal husbandry. It contributed in a major way to the economy of Faridabad thus raising its productivity, income and employment. It has land under high yielding varieties of food grains such as rice, maize, bajra and wheat. In Faridabad district, high yielding varieties of wheat were grown over 98.3 per cent of the area under wheat cultivation. The major rabi crops in Faridabad district were wheat (119.5 thousand hectares) and barley (5.1 thousand hectares) in 1989-90. While the major kharif crops grown over the same period were bajra in19.8 thousand hectares and jawar in 16.7 thousand hectares. Cultivation of fruits and vegetables covered 2,801 hectares.

Faridabad district is also successful in attracting a large number of industries, where the number of registered factories was 1,647 in 1990, of which, 1398 were working as registered factories providing employment to 1,25,862 workmen. The number of registered factories increased from 1307 in 1985 to 2045 in 1996. Faridabad town located in Faridabad district has witnessed rapid growth as an industrial town. Faridabad complex is one of the largest industrial belts in Haryana, comprising of industries in old Faridabad, Faridabad township, Ballavgarh and some surrounding villages. Such factories produce a wide range of items such as tyres, tractors, textiles, motor parts, bicycles, pottery, plastic, cloth printing, soaps, thermocol products etc. With rapid pace of industrial development, the district has various types of industrial goods to offer such as tractors, tyres, plastic toys, pharmaceutical products, chemicals, footwear, automobile parts, gas cylinders, steel casting etc. In Faridabad, the responsibility for acquiring and developing land for industrial and urban use rests with the Haryana State Industrial Development Corporation, Haryana Urban Development Authority and Directorate of Industries.

Though the industries in Faridabad have contributed to rapid industrial and urban growth they have also led to negative externalities by way of air and water pollution and irreversible damage to natural resources. The urban and peri-urban areas of Faridabad have a number of highly polluting industries such as thermal power station, hazardous chemical factories, pharmaceutical plants, steel casting plants, thermocol factories, plastic factories, textile mill, cloth mills, automobile factories, printing factory, soap factories etc. There are also polluting factories of brick kilns, in the vicinity, catering to the demand for bricks for residential and other purposes. All such factories/plants make for local air pollution by emitting/releasing industrial pollutants from Faridabad complex. Such air pollution is intensified by vehicular pollution in and around Faridabad district.

Community perspectives on agriculture and air pollution from urban and peri-urban areas form the crux of the present study. In this chapter, we present the community perspectives on agriculture, impact of air pollution, impact of industrialisation and urbanisation from 14 urban and peri-urban areas of Faridabad.

At Faridabad, field research was undertaken in 14 villages (only one area Unchagaon was from Faridabad Municipal Corporation), based on scientific sites with different types of air pollution. The names of the villages are as follows.

- Baroli
- Chandawali
- Kadhaoli
- Jhajru
- Jharsainthly
- Malerna
- Pali Kasba
- Piyala
- Sagarpur
- Sahapur Kalan
- Sahupura
- Sohtai
- Sumper
- Uncha Gaon

3.2 Farmers' Perspective on the Role of Agriculture—Faridabad

Farmers of Faridabad, in the urban and peri-urban areas under consideration, perceive agriculture to be of great significance and see a clear link between land, livestock, family and quality of life. Agriculture in most villages is the main source of livelihood, whether directly or indirectly. It is also the primary source of food grains. Perspectives of women and men from the urban and peri-urban areas of Faridabad on agriculture are as follows.

- Farming provides food, fodder and fuel to the farmers. Farming is also important for those who are landless and depend on agriculture labour market and/or for those who do share cropping.
- Though a mix of livelihood is important for having multiple access to income sources from land, livestock and labour, land is also perceived as potential collateral for future requirements.
- Agriculture also provides employment, often, throughout the year. It is generally looked upon as a buffer during periods of seasonal unemployment. As Mubina, a woman farmer of village Khandawali, Faridabad exclaimed, "we are intensely involved in wheat cultivation continuously for six months in a year. As long as we have land, we will continue to cultivate wheat" (Field notes, Faridabad villages—Meera Jayaswal).
- Vegetable cultivation is considered as a source of high income.
- Agriculture and industry are complementary and support each other. Agriculture is important to hedge against children's unemployment and also to supplement income for those with less land.

In future, there is an impelling need to carry on agricultural practices and livestock for securing livelihood. However, the farmers are of the opinion that it would be difficult to do so because of fall in size of land holdings and Government/Haryana Urban Development Authority's (HUDA) policy of land acquisition, increase in input prices and infrastructure constraints. Government/HUDA's policy of land acquisition is a major hurdle in maintaining agriculture as a livelihood. As perceived by village groups, the position of agriculture as a source of livelihood is becoming increasingly problematic. One illustration from the field is provided in box 3.1.

Box 3.1

Agro-economic and Social Changes at Khandavali (Faridabad) from the Eyes of an Agricultural Labourer

At village Khandaval there is a minority group of Dalits in majority muslim-dominated areas. Ram Lalji, a dalit while doing a time line described the socio-economic changes. "At first the chapatis made of wheat used to be soft. We also had 'Jou' and 'chana' to eat. In those fields, which had rainfall, the crop used to be around 15 maunds per acre. There were no tubewells. The yield of wheat was low. Bullocks ploughed the field. We used to work as agricultural labourers. Now in 1999, wheat is 40 maunds per acre in 5 bighas. Now, there are 3 crops grown, paddy, wheat and jowar/bajra, because of this the workload has increased. Now there is problem of water and electricity. We do not get opportunities for employment since the harvesting has got mechanised. We cannot get jobs. Whatever jobs are there are taken up by cheap labour from other regions. The agricultural and other work, which is normally at a daily rate of Rs. 70/ Rs. 80 are done by cheap labour at half the rate at Rs. 40/Rs.50. Such labour live in Jhuggis (slums) and have low cost of living."

Agricultural activities have increased and the means of agriculture have improved. Many have tractors in the village. Private factories are taking land for Rs. 2000/—per sq.yard. HUDA's rate is much cheaper though there is no land acquired by HUDA in this village. The village has factories like plastic, melting iron, machine for motor cycle parts and thermocol. The factories are fast spurting on village land, cultivable land. The farmer sells a part of the land, say 5 bighas and continues cultivating the rest of the land.

Now there is also the problem of health. Earlier, there used to be fever, small pox etc. Now, incidence of cough, cold, fever etc. has increased. Almost 90 per cent have it 3 to 4 times. There is cough due to pollution arising from smoke from the factories as per younger generation. Older generation thinks it is due to weather.

Regarding ways of life, earlier both women and men used to work hard in the field, now less people are required to work because of mechanisation.

Only 2 people are needed and the rest can enjoy leisure. People have time to raise livestock. Earlier 10 per cent use to drink alcohol and now it is 60 per cent. Dowry payment has also increased to Rupees 60,000/Rupees 70,000/-. Earlier there was no payment on account of dowry.

The workload of women has increased. The women do weeding/cleaning of fields. Their workload at home has also increased. Earlier they had mud houses and now they have pucca houses which they need to clean (mud houses required less time for cleaning). Now they have cash to buy soap to wash clothes, which takes more time. The workload of women from poor households has increased considerably. They have to go for wage employment, they need to bring fuel wood and walk for 2 to 3 kms to the forest, it takes time around 8 hours, going and coming back. They carry fuel wood, 4 to 5 days in a week. Sometimes other men loot their fuel wood saying that why do you take fuel wood from our forest.

Now we live in peace in this village In future, wage labourers will have major problem. They will not get employment opportunities. They will start looting and get engaged in dacoity/theft.

Source: Field Notes on Faridabad, Neela Mukherjee

Selected factors adversely affecting the dominant role of agriculture in urban and peri-urban areas of Faridabad are listed below.

- Sub-division and fragmentation of land has led to reduction of size of land thus affecting its viability for cultivation and modern farming.
- Acquisition of land for industrial and construction purposes is forcing residents to sell land and buy agriculture produce from markets.
- There has been increase in constraints to agricultural cultivation with new crop disease, new pest and weed and problems in procuring quality seeds and other inputs.
- Increase in family size and breaking up of joint family system is forcing family members to look for regular and stable sources of income.

- Uncertainty of irrigation, mostly due to irregular electric supply and lack of experience and knowledge act as deterrents to vegetable cultivation by local people. A sharp fall in ground water level has added to complexity.
- Rise in price of agricultural inputs has resulted in agricultural income not keeping pace with rise in cost of living.
- The younger generation is less interested in pursuing farming as a livelihood as they do not want to soil their hands. They would prefer government and private jobs to farm-related work due to stable monthly income and retirement benefits.

However, in Faridabad, agriculture as a livelihood happens to be the mainstay for hedging against uncertainties of food, employment and income. It makes for productive use of land and water and also provides fodder for livestock and often fuel wood at zero cost. On a seasonal basis it provides for a variety of items in the food basket of local communities and also contributes towards urban and peri-urban food markets. Its backward and forward linkages are of immense significance in terms of supply of inputs, marketing of produce, employment creation, income generation, processing for value addition, ware housing, trading and loaning activities.

Faridabad Villages and Agriculture

The profile of agriculture in the urban and peri-urban villages of Faridabad is briefly sketched below in box 3.2 based on descriptions provided by the local farmer groups. The box shows that agriculture is a major livelihood and also a critical one involving large sections of local communities, though its share varies from village to village.

Box 3.2
Profile of Agricullture in Faridabad Villages

- **Pali Village**—90 per cent of landed population is engaged in agriculture. Size of cultivable land has become small thus forcing the landed population to procure wheat and jowar from the market for personal consumption and for fodder respectively. A local estimate of yield of wheat crop is around 12 to 13 quintals per acre.
- **Sohtai**—Farming activities involve 75 per cent of the local community though 80 per cent of income accrues from agriculture.
- **Sahapur Kalan**—In this village livelihood of people depends on land, livestock and salaried job with 70 per cent involved in farming and livestock, 20 per cent in service sector jobs and 10 per cent in labour work.
- **Baroli Village**—Share of population in agriculture has fallen where 50 per cent of the population is engaged. Fragmentation has resulted in small sized land holdings with the yield of wheat crop estimated around 16 quintals per acre. Cash crops prove uneconomical on small lands. Vegetable cultivation suffers due to constraint of irrigation water. There is a sharp fall in ground water level and irregular electric supply. Breakdown of joint family system has forced people to look for alternate sources of income. Milk business is lucrative due to proximity to urban areas.
- **Jhajru Village**—Agriculture is still the most important occupation though acquisition of land has led to decline in area under cultivation. The yield of wheat crop is estimated to be 16 quintals per acre.
- **Piyala Village**—Its main occupation is agriculture. The soil is salty and not good for cash crop cultivation. Increasing input cost is also making agriculture less profitable.
- **Jharsainthly**—Around 20 per cent of the land has been kept for cultivation, while 70 per cent of the village land is under Haryana Urban Development Authority. The cultivable land is affected by alkalinity hence, its crop potential stands limited. Wheat is grown during Rabi season while most of it is kept fallow during other seasons.
- **Sahupura**—Earlier dependency on land and livestock was high but now, nearly 50 per cent of the community members work in the nearest town of Ballavgarh and the rest do farming, of which, 10 per

cent do share-cropping to sustain their livelihood.

- **Kadhaoli Village**—Agriculture is the main occupation for almost 75 per cent of villagers though sharply declining over the years. An average estimate of wheat crop is around 13 to 15 quintals per acre.
- **Sumper**—Livelihood-related activities constitute both farming and non-farming activities where 50 per cent of the population depend on farming and the rest depend on non-farming, which includes 5 per cent regular job, 10 per cent business and 35 per cent labour work. Share cropping is not very common in agriculture.
- **Chandawali**—Livelihood-related activities depend on both agriculture and non-agriculture. Village income pattern shows that 60 per cent of income is from agriculture and 40 per cent from service sector jobs. Haryana Urban Development Authority has acquired about 200 acre of agricultural land while negotiations for further acquisition is continuing.
- **Malerna**—In this village, the villagers depend on farming-related activities to the extent of 50 per cent and the rest is on labour in the local industries. A local estimate of yield of wheat crop is 2600 kg per acre. Earlier, farming and livestock made for 80 per cent of livelihood-related activities of the villagers while artisan skills constituted 20 per cent.
- **Sagarpur Village**—Agriculture is an additional source of income; it is a part time-occupation. An estimate of yield of wheat crop is 14 quintals per acre. Around 95 per cent of villagers is engaged in selling milk and 60 per cent is engaged in agriculture.
- **Unchagaon Village**—There is uncertainty in pursuing agriculture as a livelihood due to recent acquisition of land by Haryana Urban Development Authority. A rough estimate of yield of wheat crop is 16 quintals per acre.

Source: Based on field reports of PA researchers on Faridabad

As described by the farmers of urban and peri urban villages of Faridabad, three types of crops are grown, kharif, rabi and jayad. Since the 1970s, wheat is the main rabi crop (for consumption) and jowar is the main kharif crop (for fodder). Paddy is also grown in some of the fields. Earlier, in the 1940s and the 1950s coarse grains (barley, bajra) and pulses (gram) were the major crops. From the late 1980s, a variety of vegetables are being grown round the year in those villages where water is not much of a constraint.

There has been a clear shift in cropping pattern from sama, gram, barley, bajra, sugarcane and lintel to wheat, jowar, paddy, mustard and vegetables. New crops give higher yield and better market price. Many farmers have substituted improved and high yielding varieties in place of low yielding varieties and coarse grains.

Livelihood and Coping Strategies at Faridabad

As described by the local communities, most household members are engaged in multiple livelihood—related activities, as return from any one activity is not sufficient. The livelihood-related activities are listed in boxes 3.3 and 3.4.

Patterns of Livelihood

As shown in box 3.3, men are engaged in various activities and some of the senior ones receive government pension. Few have also migrated to Gulf countries for work for e.g. from Kadhaoli village. The work done by wage labourer relates to agricultural sowing/harvesting in Rabi and Kharif season and non-agricultural work during rest of the year.

Women from local communities, in general, are not allowed to move out in search of work and are mainly involved in farming and animal husbandry. Some of them earn income, mostly in kind, while working as agricultural wage labourers. Others are engaged in their own field (not earning income), and/or in animal husbandry within their households. The women are also involved in home making, collecting fodder, fuel wood, milk etc. and some of them also work as helpers in Anganwadi.

Men from each household are engaged in different occupations including agriculture while women do their traditional work as above. Many households are engaged in the milk business, which gives them additional back up in terms of income. In a village like Pali Kasba, local mining activities have provided additional source of income.

Box 3.3
Livelihood-related Activities in Villages of Faridabad

Men
- cultivation of land
- livestock rearing
- milk selling
- agricultural labourers
- wage labourers in cities
- workers in tempo trucks
- workers in tourist vans
- cart pulling
- doing petty trade
- running shops
- working as rural artisans including pottering
- barbers
- carpenters
- goldsmith
- weavers
- government service
- private service
- doing mining-related activities
- others

Box 3.4
Work done by Women in Faridabad Villages

- land cultivation
- livestock rearing
- housekeeping
- collecting fodder, fuel wood, milk etc.
- few working in anganwadi

Source: Field Reports from Faridabad of PA Researchers

The incidence of indebtedness varies across villages with some villages like Khadaoli have a high level of indebtedness while village Pali Kasba having a relatively lower level. The local community members take loan for different purposes such as marriage, house construction and renovation, medical treatment, buying/hiring tractor, purchasing buffalo, procuring agricultural inputs like seeds, fertilisers, pesticides etc. The loans are contracted from different sources such as relatives, friends, businessmen, moneylenders, mortgage bank, Ballabgarh, regional rural banks etc. Such loans taken by households are spread across different sources and are not generally from a single lender.

3.3 Farming Practices and Seasonal Activity Calendar

As described by the women and men groups of Faridabad villages, three types of crops are grown, kharif, rabi and jayad. Since the 1970s, wheat is the main rabi crop (for consumption) and jowar is the main kharif crop (for fodder). Paddy is also grown in some of the fields. Earlier, in the 40's and 50's coarse grains (barley, bajra) and pulses (gram) were the major crops. From the late 80's, a variety of vegetables are grown round the year in those villages where water is easily available.

There has taken place a clear shift in cropping pattern from sama, gram, barley, bajra, sugarcane and lintel to wheat, jowar, paddy, mustard and vegetables. New crops give higher yield and better market price. There has been improvement in kinds of crops grown where better crops have been substituted for coarse grains. Also the farming community has stopped growing grams since the seventies because of high moisture content of the soil due to use of tube well water. Decision-making by farmers on the kind of crop to be cultivated is determined by 6 factors.

- water requirement
- land size
- experience
- time required
- costs of cultivation and
- returns on crops

The farmers in the villages of Faridabad under reference use tractors for ploughing fields, sowing seeds, for irrigation and harvesting of crops. Labourers are used by small farmers for planting paddy saplings, vegetables like cauliflower, brinjal etc. and for harvesting of paddy and wheat. In general, in an acre of wheat, ten labourers are employed and they harvest the field in one day's time. The labourers, as a group, are paid around 4 quintals of wheat for distribution amongst themselves. Fields are irrigated through tube wells run on electricity. Hybrid seeds, pesticides and fertilisers are used as inputs by the farmers though no proper training and guidelines have been provided to the farmers by the agriculture office regarding appropriate use of inputs.

The farmers run high risk with new/unknown pest/disease/weed attack and apply measures on their own to treat such pest/disease/weed. If successful, mutual extension of knowledge is undertaken. Farmers appear to be risk-averse in sowing new hybrid seeds available in the market. Only 3.5 per cent of the farmers take new initiative while other farmers follow in case of good results. Some farmers got their soil tested in earlier periods while now the process of soil testing has become time-consuming and not that easy. According to the farmers their quality of land has deteriorated with the use of fertilisers and pesticides. If wheat crop is taken as the basis for comparing crop yield over time, then such yield has either become stagnant or has marginally declined over time. There is also drop in the size of wheat grain and its taste accompanied by loss of colour, though mixed reasons were cited for such trend.

The seasonal calendar, as given in table 3.1 shows that the farmers are busy throughout the year in growing different kinds of crops and vegetables. Such agricultural activities cover a broad spectrum, preparation of land and sowing of crops/ vegetables to selling of products in the market. Apart from supporting the usual agricultural activities, the women are also required to arrange fodder for their livestock. Milk from livestock, a lucrative source of livelihood, is so important that one major fodder crop— jowar is grown during May-September along with paddy and different vegetables. The months from November to May are devoted to growing wheat, another major crop, a part of which is also used as fodder. The seasonal calendar shows the range of vegetables, which are cultivated by the farmers, especially in the winter.

Table 3.1   Seasonal Calendar: Activities Related to Agriculture— Faridabad

Month   **Description of Agricultural Activity**

Jeth-Asar (15 May-15 July)
Preparing land after cultivation of barsam and jai; sowing of jowar, paddy, bajra, arhar dhencha (fodder crop), moong, sesame, kakri (cucumber-type), til, moat, maize; cutting of some jowar for fodder; sowing of gowar, bottle gourd, chilly etc.; selling some vegetables sown in Baisakh; irrigating vegetable field; selling lobia, karela, cucumber, onion, lady's finger, bottle gourd.

Sawan (15 July-15 August) Uprooting wheat from fields; arranging fodder (women); sowing of dhencha; cutting of jowar for fodder; some farmers preparing samples of brinjals and cauli flower; irrigating land, if there is no rain, and if electricity is available; spraying pesticides in paddy fields for killing snakes and other pests; planting paddy in Sawan; providing fertilisers to paddy fields; planting saplings of spinach and cauli flowers; selling cauliflower, bottle gourd and bitter gourd, onion and lady's finger; some farmers spraying medicine for locusts; weeding fields, if required and if possible.

Bhado (15 August—15 September)
Irrigating paddy and jowar fields, arranging fodder (by women); cutting of dhencha and jowar leaves for fodder (by women); planting cauliflower, potato and raddish; ripening of fruits on jowar crops; stocking of some jowar seeds for use in next crop cycle; using the rest for personal consumption and sale.

Kuwar-Katak (15 September-15 November)
Harvesting jowar, bajra and paddy; preparing land for wheat; applying water and manure on field; sowing of wheat, barsam (fodder crop); cutting of some jowar for fodder (by women); sowing seeds of radish, carrot, spinach, cabbage; planting saplings of cauliflower and brinjal in the field; sowing of barsam and jai in Katak; harvesting bajra in Kuwar and keeping it for 3-4 days for drying; at the end of Kuwar, using bajra for making chappatis.

Aghan-Pus-Mah-Phagun (15 November to 15 March)
Applying several rounds of irrigation of wheat fields and arranging of fodder (by women); spraying of fertlisers to fields; in Aghan, sowing of wheat, mustard, cauliflower, karela, cucumber, lobia, turai, bitter gourd, spinach, methi, carrot, raddish and potatoes; irrigating vegetable fields; spraying fertilisers; looking after

ear rings in barley; harvesting and selling vegetables such as carrots, cauliflower, lobia, brinjal, lady's finger, gourd, gowar, potato, cabbage, spinach, chillies; also weeding wheat field; collecting grass and barsam leaves for fodder (by women); harvesting of mustard in Phagun; big farmers selling surplus mustard after storing some for personal use, small farmers extracting oil from mustard for personal use; using the extracted waste of mustard as animal feed and collecting seeds for further processing.

Chet-Baisakh (15 March-15 May)
Harvesting wheat, threshing, storing, drying and selling it; collecting fodder (by women); also irrigating fields; selling brinjal, onion, melon, lobia, turai, chillie, pumpkin, bitter gourd, spinach; cutting of barsam and white jai for fodder, as required; in Chet, preparing field for kharif crop; tilling, ploughing, giving manure and putting water to land.

Note: This is a general seasonal calendar as described by women groups and men groups from selected villages of Faridabad. Not all households of the villages are engaged in all activities mentioned above. There are different crops and vegetables sown in different villages and the above table provides a general picture of agricultural activities undertaken in the villages of Faridabad, under study. For village-specific activity see field results related to the villages

Source: Based on field reports of PA researchers on villages in Faridabad

3.4 Agricultural Problems/Constraints

Local Communities' Analysis of Problems/Constraints

The generic problems and its local dimensions as described by the local communities at Faridabad villages have been assembled in box 3.5. The problems have been listed in six generic clusters, which are weather-related, weed-related, pest and insect-related, disease-related, animal-related and infrastructure/input-related. All such problem clusters act in their own ways in constraining growth of agricultural crops, yield and production. However, some problems are perceived to have greater impact than others and are also seasonal in nature. In this section, the problem clusters have been listed, their seasonality factor has been described and they have been prioritised as per community perspective.

Box 3.5
Issues within Problem-Clusters in Agriculture—Faridabad

- **Weather Condition-related**—hailstorm affects all crops and damages them; its maximum damage is on crops in their infant stage; unseasonal and heavy rain accelerates the growth of weeds amongst crops like wheat, mustard, jowar and bajra at any stage; excess rain damages the standing crop; lightning damages mustard flower.

- **Weed-related**—Sati is a variety of local grass; it arrests the crop growth of jowar, arhar, vegetables and paddy; it originates from fertiliser; mostly active at the infant stage of the crop, competes with the crop for nutrients; absorbs all the fertilisers; rise in incidence of sati in the last 10-12 years; however, if it rains before the crop grows then sati dies and has minimal impact; uprooting the weed is quite difficult; it is used as fodder; Katoll—affects wheat crop and stunts its growth; it is quite difficult to uproot it; it is used as fodder and does not create health problem; Bhatua—stunts the growth of wheat crop at any stage; Congress grass/ gajar grass (Parthenium)—stunts the growth of crops like wheat, jowar, bajra, arhar and mustard; it can occur at any stage; it grows faster than the crop and turns into bushes; it creates smell and allergy and can lead to respiratory problems; it came with Australian wheat grant in the 1970s; it grows throughout the year and affects all the crops; it causes allergy and itching; it can host a number of poisonous insects in its bush; for the last ten years it has become rampant; once uprooted it re-grows quickly; incidence of safed moondi increased in recent years; affects wheat crop, it competes with wheat for nutrients mostly at the infant stage; its impact is lessened by uprooting; rani jai—increased incidence of rani jai, which stunts the growth of wheat mostly at the infant stage; it has to be uprooted at regular intervals to protect the crop; competes with wheat crop for nutrient; increased incidence over the last 8 years; mundi, gehun ka mama, kanki (Phararis Minor)—commonly found in wheat field and has close resemblance to wheat crop; it sucks the nutrients from the field and overpowers wheat crop and can lead to massive loss in crop yield; the local view is that such grass came with hybrid Mexican seed of wheat; no effective measures available against parthenium (congress grass in all fields), rani jai (in wheat), sati (more in millets); safed moondi (in wheat) etc.

- **Pest and insect-related**—Chepa occurs naturally and damages crops e.g. mustard and bajra, mostly at the youth stage; leaves of the crops becomes sticky and dries; covers and eats the mustard flower and dries the flower; Sundi pest sticks to the leaves of paddy and sucks the juice out of it; Kanduwa attacks matured wheat; Geddar attacks young vegetables; Lairri eats the leaves of vegetables; Locust attack increasing for the last 3-4 years (first seen 4 years back) in the kharif season; eats leaves of jowar, bajra and dhencha crops at stages of growth; aided by dry weather; termite—eats up roots of wheat, jowar and bajra, mostly occur due to inadequate rainfall and improper irrigation; grasshopper—comes in herds and attack matured crop fields, destroying kharif crops; their size has increased in the last ten years, which is attributed to the improved variety of seeds being cultivated; other small insects—different variety of small insects are found at the root, on the leaves, stem, twigs of crops; some eat away tender leaves, flower and thus disallow grain formation; their incidence has increased over the last ten years, which is attributed to growing use of fertilisers, high yielding seeds and cropping patterns.
- **Diseases and pollution-related**—Yellowing of leaves is a common occurrence in vegetables and wheat; disease noticeable since last 10 to 20 years; leading to pre-mature drying of plant; affecting crop growth; reasons for such disease attributed to factors like smoke and dust from near-by factories/due to use of chemical fertilisers and pesticides in large quantities; curling of leaf (Blight)—affects potato, chili and brinjal; leads to twisting of leaf and affects yield; prevalent in improved variety of crop and its reason attributed to problem in quality of seed and pollution from factories; blast disease affects paddy crop, especially Musthar Basmati; the local people fear that the disease can spread to other crops; in this disease, different parts of the plant dry up and crop gets destroyed; leprosy affects sugarcane crop, its joints and dries up the plant; this disease is occurring with improved variety of sugarcane; smoke from near-by thermocol factory gets deposited on the vegetables like cauliflower, spinach etc.
- **Animal-related problem**—Increased invasion by Nilgai, which is an endangered species protected under wild life protection; they move in herds of 10 to 15; they destroy the crop by trampling and sitting on it; grazes on the field of jowar, bajra, wheat and mustard and destroys crop; they enter the mustard field the most for it makes them feel warm; they also destroy the flower and premature crops of lintel, peas and sugarcane; this had made some farmers stop growing the crops;

the attack is more when the crop has matured; wheat crop is hard for them to consume and thus gets saved from their attack; stray cow come in herds and destroy standing crops of wheat and vegetables; field rat is generally problematic for wheat and paddy; eats grains and bores holes in the fields; can attack the crop at any stage; also monkeys are around to damage crops.

- **Infrastructure/Input/Output-related**—erratic supply of electricity, tube wells cannot be operated for irrigation; unable to irrigate land; this is due to power being supplied to industry; forced to depend on rain and tractors for irrigation; water is hard in the past 5 years, which leads to crop getting wasted; hardness has increased due to boring of electric polls by National Thermal Power Corporation 4-5 years back; ground water level falling rapidly; little or no water in government canal; using water from sewage canal; acquisition of agricultural land by HUDA is impacting on their livelihood; fertiliser not released in time by the government and hence problem of crop not getting nutrients in time; non-availability of good quality seeds in time; high difference in buying price of input and selling price of output; in winter, the cold nights and thick fog pose problems in doing irrigation of fields.

Note: As described by women groups and men groups from selected villages of Faridabad

Source: Based on field reports of PA researchers on Faridabad

Seasonality of Agriculture Problems/Constraints

One important dimension of agriculture problems/constraints at Faridabad is that they are seasonal in nature. The seasonality aspects are indicated in table 3.2. The table provides an idea of those seasons when the problems/constraints become acute based on which interventions for minimising or overcoming the problems can be planned.

Table 3.2   Agriculture Problems/Constraints/Risks in Faridabad

Jeth-Asar (15 May-15 July)
Weather/Climate-related—In Jeth, hot winds damage creeper crops like cucumber, tori, melon, bottle guard etc.; in Asar, late monsoon results in crop damage (upto 25 per cent); excess rain results in crop damage (as much as 40 per cent to 50 per cent); rains can damage harvested wheat; sudden rains can damage the seeds sown since it hardens the upper layer.
Weed-related—Motha (weed grass) affects crop of cucumber and kakri (impact increased in last 7-8 years); congress grass inhibit jowar crops; sai weed attack the infant jowar crops; baroo (weeds) and sati (weeds) stunt the growth of jowar; sati (its visibility increased in the last 5-6 years).
Pest and Insect-related—Gendar (caterpillar) attacks roots of the crop/s and dries it; locusts feed on Jowar leaves for the past 3 years (crop damage can be as much as 80 per cent); sundi attack in vegetables (if hot winds blow then sundi automatically dies); chepa attacks on bajra for the last 3 years; chitli damages the roots of the young vegetables; termite attack the infant jowar crops; locust attack the young jowar crops; locusts and termites damage jowar and barsam crops in their early stages.
Disease-related—Moriya (type of disease) in chilies (at matured stage) (This year 80 per cent crop affected, reason unknown for the disease).
Animal-related—Nilgai destroys jowar and bajra crops; monkeys damage jowar and barsam crops in their early stages; stray cows eat and destroy young crops.
Infrastructure/input-related—In Asar, lack of water for irrigation; problems in irrigation due to irregular electric power supply; high cost of inputs.

Sawan-Bhado (15 July-15 September)
Weather/Climate-related—Those as above in Jeth-Asar plus danger of excess rain damaging crops (almost by 50 per cent), saplings; rain water washes away the pesticides.
Weed-related—damage by bajal weed and gander affecting paddy fields and brinjal saplings; congress grass arrests growth of crops and causes allergy in humans.
Pest and Insect-related—sundi pest get stuck to the paddy leaves and sucks the juice out of it; sati prevents jowar seeds from germinating; sati arrests growth of jowar crops; chepa (pest) attack on bajra leaves affecting size and quality of its grains; taking them as fodder can make buffaloes sick; locust attacks on young leaves of jowar and bajra (damage upto 80 per cent); mosquitoes eat the leaves of green leafy vegetables like spinach.

Disease-related—cauliflower saplings dying by itself (this year about 80 per cent to 100 per cent died).

Animal-related—damage by nilgai to jowar and bajra crops (as much as 50 per cent).

Infrastructure/input-related—pesticide remains in the crops unless washed away by rains, which is poisonous; for the last two years pesticide spray very expensive and ineffective, if not used simultaneously by all the farmers; high cost of inputs; irregular supply of electricity.

Kuwar-Katak (15 September-15 November)

Weather/Climate-related—In Kuwar, excess rain damage the standing crop/s; there is danger of frost where damage can be controlled if water is available for immediate irrigation.

Weed-related—weeds like rani jai and safed moondi affect wheat crops' growth, no medicine available for rani jai, if uprooted, the wheat crop also stand uprooted; when not uprooted, it grows more; for safed moondi, some medicine is given, which proves 50 per cent effective though fake bottles are also available; kana (fungus) affects inner portion of vegetables like cauliflower.

Pest and Insect-related—sundi (pest) affects vegetables; titris (pest) can damage crops; bees can damage crops.

Animal-related—nilgai eats the crop; monkeys damage crops; stray cows damage the crops; rats damage standing crops.

Aghan-Fus-Mah-Phagun (15 November-15 March)

Weather/Climate-related—in Mah, crop damage due to hailstorm (as much as 40 per cent); sudden rain can damage sown seeds and fertilisers applied to land; in Phagun, danger of wind damaging wheat crop also exists; crop damage by electric power cuts and hail storms (as much as 20 per cent); lightning can damage the mustard flower and lowers the yield; farmers have to irrigate land in cold nights; sometimes wind damages the wheat crop and weakens crops; high cost of inputs; irregular supply of electricity.

Weed-related—Damage due to jai weeds in wheat fields (as much as 20 per cent to 30 per cent); black jai and safed moondi reduces the wheat crop by disallowing wheat seeds to germinate; rani jai and safed moondi suppresses growth of wheat crops; rani jai (damage to crops by 50 per cent), safed moondi (damage to crops by 75 per cent) and other wild plants (damage to crops by 20 per cent) like morila, kateli, bathua impedes the growth of crops especially wheat.

Pest and Insect-related—gandar attack on young vegetables; lairri feeds on the leaves of young vegetables; mustard flowers are affected by chepa; chepa covers and eats mustard flower, which dries as a result; sundi (pest) in the vegetables; in Phagun, rust affects wheat; kanduwa affects the wheat and no treatment is available; chepa or pharka (damage to crops by 40 per cent) on mustard due to moisture in air, more in foggy weather; uksa affects the earrings of the wheat, seen more in sandy soil (10 per cent loss due to this); In Aghan, crop damage due to termite (as much as 25 per cent); threat of termite attack if water not given for 20-25 days.

Animal-related—crop damage by nilgai especially mustard; crop damage from nilgai; rats make holes in the field and weaken the roots of the crops; nilgai, stray cow, and monkeys damage crops.

Chet-Baisakh (15 March-15 May)

Weather/Climate-related—Crop damage due to un-seasonal rains (as much as 20 per cent to 30 per cent); excess rain, frost (can damage as much as 90 per cent as in potatoes last year) and strong wind damage wheat crop; hailstorm and rain can damage harvest crops left drying in the field; rain can result in kanduwa on grams (virus) if the grains are not properly stored.

Weed-related—lalrri eat the leaves of the young vegetables; danger from sureti (pest) and pal (pest) on stored grains in case of wet weather (though pesticide against it is effective).

Pest and Insect-related—white and black kanduwa on matured wheat with crop damage (as much as 5 per cent).

Animal-related—monkeys also eat and damage the crops; stray cows eat and damage crops.

Infrastructure/input-related—pesticide used but not effective; poor farmers cannot afford adequate fertiliser and pesticide.

Note: As described by women groups and men groups from selected villages of Faridabad

Source: Based on field reports of PA researchers on Faridabad

Prioritised Problem Index of Faridabad Communities

The next issue to be analysed is whether all problems/constraints in agriculture are equally important or whether some are more important than others. The issue is about the way in which the problem/constraint clusters were prioritised by the women and men group/s of Faridabad villages. The listing, scoring and ranking of problems/constraints in agriculture as done by different women and men group/s have been assembled for inter-community comparisons based on elementary statistical tool of indexing. Some groups used fixed scoring (i.e. expressing priority through apportioning a fixed number of seeds/stones to the problems), other used free scoring (using any number of stones/seeds/etc. to score), others only ranked problems without scoring and some groups listed their problems without scoring or listing. Hence, there emerged a range of ways in which groups did their problem prioritisation. For comparison across communities in Faridabad, the scores, ranks and frequencies of problem listing have been indexed to arrive at prioritised problem indices (PPI) of agricultural constraints/problems as shown in table 3.3 and bar chart 3.1. This has helped in retaining diversity of problems and maximum information, which local communities from Faridabad have provided. Such PPIs have been constructed for women and men groups separately.

Table 3.3 Prioritised Problem Indices (PPI's) of Agricultural Constraints/Problems—Faridabad

Problem/Constraint Criteria	Rank by Women	Rank by Men
Weather/Climate and natural resource conditions-related	3	6
Weed-related	2	3
Pest and Insect-related	1	2
Animal-related	4	4
Disease and Pollution-related	6	5
Infrastructure/Input-related	5	1
Organization-related	—	7

Note: The above is based on the prioritised problem index of the village community resulting from listing, scoring and ranking exercises done by local groups of women and men in Faridabad villages. The various groups managed the process of prioritising problems differently, while some used problem ranking, others did problem scoring and still others merely listed their problems. For problem scoring, many groups used seeds or stones to indicate their priorities (with greater the number of seeds greater the priority). For other details, see methodological note

Source: Based on field reports of PA researchers on Faridabad

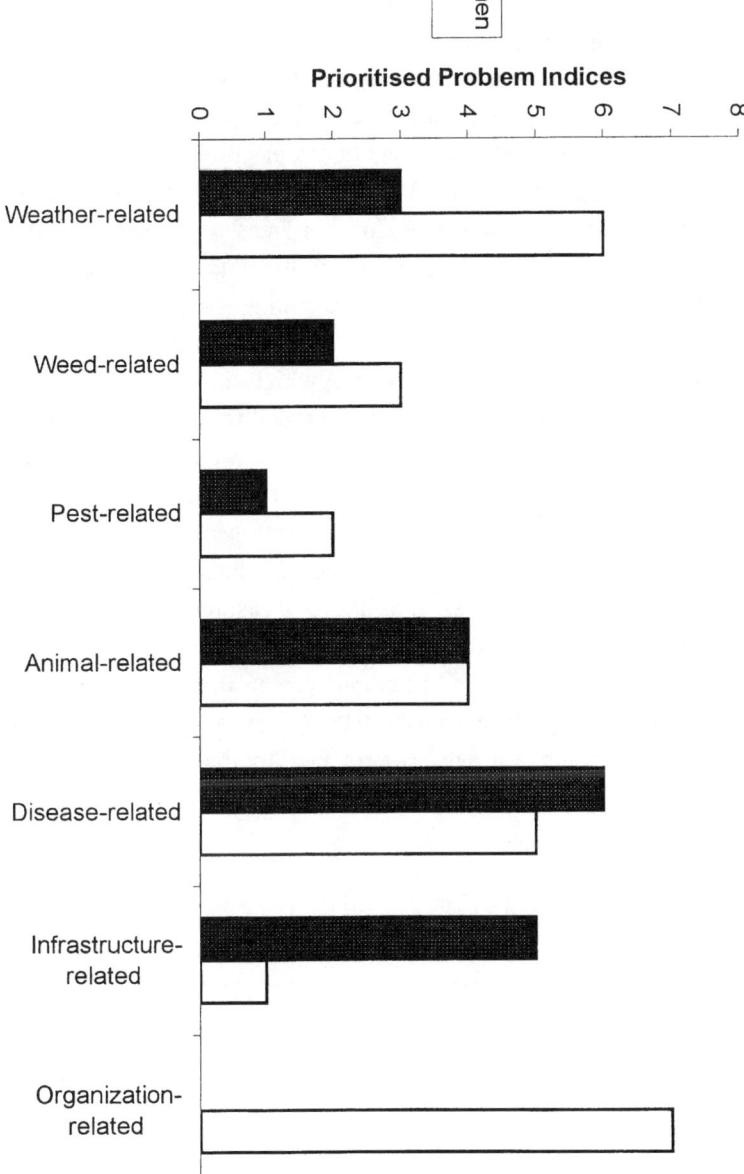

Gender perspectives on Agriculture Problems/Constraints

As the PPI in table and chart 3.2 shows, pests and insects have been perceived by the women groups of Faridabad as the foremost constraint which men group consider as their second problem following problems and constraints related to infrastructure and inputs in agriculture. For women groups of Faridabad villages their work in the agriculture field make the problem of pest appear most important followed by the problem of weed. Some of the common pests the women consider most problematic are locust, termite and chepa while the four weeds, which pose as common constraints as perceived by women are sati, congress grass, safed moondi and rani jai. For men groups of Faridabad villages, infrastructural constraints are of prime concern. Such men going for procuring seeds, fertilisers, pesticides etc. are most concerned about the falling standards of agricultural inputs, erratic electric supply etc., which they think are crucial to agriculture. The problems of pest, insect and weed follow the problem of infrastructure and inputs. Again the problem of weather and natural resource conditions is the third priority problem of women groups while it has a much lower priority for men, being their sixth problem. The women perceive weather conditions as their third problem. This has mostly to do with sudden hailstorms, frost, less rainfall, un-seasonal rain, which do extensive damage to crops and adversely affect livelihood related to agriculture. For both the groups, wild animals are an important part of the problems concerning agriculture for they lead to loss of crops and the farmers are helpless for such animals are protected by the Wild Life Act.

Coping Strategies

In the urban and peri-urban villages under study in Faridabad, some of the problems/constraints were most often overcome by personal efforts. These include the following.

- The farmers substitute the vulnerable crop variety like chickpeas, lintel and sugarcane so as to overcome selected agricultural constraints like damage from pest, weed and animal.

- For preventing nilgai from damaging crops, the farmers adopt short and long term measures like putting barbed wire around the field, if possible; making noise to scare them; staying in the fields at night to guard the fields; etc.

- The farmers tried cordoning off stray cows, which damaged their fields. However, this is effective when the owner comes and apologizes. Or alternatively, if the owner is identified, the cows are restored back with a warning to the owner. In other cases, the owners of the cows from near-by towns have registered police cases against the farmers involved in keeping the cows as captives.

- Sometimes rat poison is given to kill rats or rat trap is used.

- Some farmers also inform the local agriculture office about pest attack and weed growth, though they seldom find it worthwhile to spend time in finding ways to overcome the agricultural constraints. This is because of their small land size and engagement in other livelihood-related activities, which keep them busy.

- Since agriculture has become less profitable due to diminished land size, increased pest attack, weed growth and weather fluctuations, the younger generation has shown strong preferences for other kinds of livelihood.

- In order to cope with falling returns from agriculture, the local community members look for alternative sources of income in and around the village. In this context some farmers whose land has been acquired by the government have decided to use the compensation money to buy new land in interior villages and increase their farm size (one e.g. is that of village Unchagaon).

- In village Sagarpur, the residents are seriously considering selling of land to private parties (especially industrial firms) before the government acquires such land. This is because of two reasons, one is that the price of land realised is higher from private parties and the other is that the government policy makes it mandatory for industrial firms to provide job to a member of household from which such land has been purchased.

- Community members of village Sagarpur also stated that they were developing a recreation park in which around 30,000 trees of

Sesama, Neem, Banyan, Mahua etc. have been planted to reduce the air pollution.

- In a village like Sagarpur, the local community members appealed to the local development authority to check the mysterious death of their buffaloes for the last two years following which some visits by the authority were made to the village though not much could happen.
- Regular application of pesticide is done to guard against pests and insects.
- Weeding is done on a regular basis.
- In the crisis months, the local community members also resort to seasonal migration to different places.

Some Village-specific Findings

Some village-specific findings are listed in box 3.6 for having an idea of the problem/constraint at the locality level.

Box 3.6
Cluster of Agricultural Problems and Constraints, Faridabad

Weather Conditions-related
- In Sahupura, one women group pointed out hail storm as one primary factor causing less cultivation of mustard and laid down paddy thus reducing its yield from 70-80man/kila (recorded last year) to 25 man/kila this year; both women and men group/s pointed out that thick fog in the current year made spraying of pesticide and weedicide ineffective.

Weed-related
- In Malerna, complaint of parthenium (Congress grass) is high. In the last three years this weed has spread in the whole area and has also affected the crops in the fields.
- In Sahapur Kalan, weeds are spreading faster than before.

Pest and Insect-related
- In Sahupura, white ant is quite common in the fields because of dry soil and they destroy the plants by their roots; pest attacks on paddy crop from caterpillar and chepa increased due to adverse weather conditions.

- In Sahapur Kalan, some variety of paddy and wheat are more vulnerable to pest attack.
- In Chandawali, incidence of pest and disease has increased with the introduction of new variety of seeds.

Animal-related

- In Sahupura, stray cows from Ballavgarh and blue bull attack their fields at night.
- In Sahapur Kalan, blue bull destroyed young plants at the flowering stage.

Disease and Pollution-related

- In Sumper, most of the crop diseases are new and are prevalent in improved variety of seeds.
- In Sohtai, there is pollution from brick furnace which reduce the yield.
- In Sahapur Kalan, plant diseases are considered most important problem with yellowing of crops, blackening of leaves and curling of leaves as common occurrence.
- In Chandawali, potato crops, blight disease is common this year in the improved variety, Kufri Anand.
- In Chandawali, large quantities of vegetable crops get spoilt due to pollution.

Infrastructure/Input-related

- In both Sumper and Malerna, irregular electricity supply disturbs irrigation facility. Often, villagers spend nights in the field to irrigate their land. Unadulterated fertiliser and pesticides are not available on time and adulterated inputs are not effective.
- In Sahupura, there is no timely supply of fertiliser and pesticide.
- In Sahupura, increased dose of pesticide increases cost and also sometimes damages crops due to its poor quality; farmers are forced to buy low quality inputs from private shops; forced to pay water fee to government despite getting no water, also paying rent to tube well owner for irrigating the fields.
- In Sohtai, increased price of agricultural inputs creates problem for farmers.

Note: As described by women groups and men groups from selected villages of Faridabad

Source: Field Reports of PA Researchers from Faridabad

3.5 Impacts of Air Pollution at Faridabad

Different kinds of pollution are prevalent in Faridabad as described by the local village communities under research. Such pollution is basically related to air and water. Air-pollution is linked to industrial estate in near-by towns of Ballavgarh and Faridabad, vehicular-related, local factory-related. Ballavgarh and Faridabad industrial belts are close to the villages and the smoke and different pollutants from the factories pollute the air and settle on crops and affect crop yield. Box 3.7 provides a woman's perspective on clean air.

Box 3.7
What is Clean Air?

"The clean air", Rajwati of village Sagarpur (Faridabad) said, "has no smell, no dust and no smoke". She further said, "Now-a-days, everywhere there is unclean air". Then I requested the women group to tell how air became unclean. They narrated the following sources, which causes clean air to become dirty or unclean.

- smoke emitted from the vehicles;
- smoke emitted from the factories;
- smoke and dust emitted from the brick kiln;
- drains and garbage.

Source: Meera Jayaswal's Report on Faridabad villages

The air pollution has increased due to considerable increase in the number of vehicles in that area over the last 20 years. Some of the impacts are listed in table 3.4. The pollution has led to smoke and dust in the air; it has also affected crop production. Pollution through local factories and industrial units like those of brick kilns, thermocol, plastic and cement pipe factories have affected crops and health of both human beings and animals. Polluted water from industrial units, found in some villages has affected crops.

It has also affected health of human beings and animals. The water of Agra canal got polluted about one and a half year back near village Sahapur Kalan. This adversely affected crop and vegetable cultivation, both taste and nutrition level, as described by elderly farmers during field research.

Table 3.4 Pollution—Sources and Impacts, Faridabad

Name of Village	Possible Impact Identified	Selected Observations and Present Measure/Action
Malerna	• Adverse Impact of local thermocol factory smoke is perceived in this village; the smoke is intense when waste industrial products are burnt every 15 days on the road side; the smoke deposit on leaves is visible; when livestock consumes affected fodder their milk starts smelling. • Impact of smoke on health leads to asthma and T.B. • In the absence of electricity, when the factory runs on generator, it creates high level of noise pollution.	• Complaints about noise pollution to higher-ups in local administration and the factory owners have not helped. • There was strong resistance for giving permission to the local factories that emit pollution. No consultation took place with local people for establishment of such factories and there is no forum to listen to their complaint.
Sahupura	• Mango and guava orchards are being destroyed by pollution; it was reported by local community members, that 50 years back they got 500 k.g. from 1 acre, but now with increased inputs they get 900 k.g. • Pollution from near-by factories at Ballavgarh lead to diseases like spots and patches on fruits and vegetables, black and brown patches on leafy vegetables (cauliflower, spinach, methi, raddish etc.) and yellow patches on wheat leaves. Such impact felt for the last 7-8 years. 75 per cent loss reported in vegetables like cauliflower, spinach and methi.	• For reduced yield, the older generation blames the wind blowing from the direction of the factory at Ballavgarh. It is also due to increase of no. of vehicles leading to vehicular pollution, which leaves a thin layer of dust and smoke on the leaves which inhibits crop growth and results in crop decay and loss. • Some farmers have decided not to grow brinjal from next year due to "kodia" disease and "yellowing" of leaves.

- Some farmer reported brinjal crops getting damaged (seen earlier also but now increased) increasingly over the last 3 to 4 years; 35 per cent of such loss was due to "Kodia" disease (in which the brinjal looks deformed). Some microbes in the roots (as told to farmers by local scientist from Ballabgarh) apparently cause it. There is no measure except uprooting it and burning it since it is infectious.
- In the current year, for the first time, "yellowing" of top leaves was observed in 50 per cent of the crops. The farmers cited no reason where such problem was observed.
- During the time of ripening of paddy, the leaves started wilting in many fields, which lowered the output. Titri (pest) also attacked paddy at the time of its ripening (Kuar end) and sucked the milk out of raw grains.
- Farmers almost stopped growing mustard due to air pollution, which damages flowers (though some community members opined that chepa, a type of pest can be equally responsible).
- Water pollution due to acid flowing through canal water from the industrial area burns the crop near the canal until the end where the water reaches. Such impact being felt for the last 8-10 years.

- Air pollution corrodes the surface of some vegetables e.g. chilies.

Khadaoli
- The fumes emitted from the factory mixed with dew in winter,

especially in January and February falls on wheat and gets deposited. If this continues for a month the size of wheat grain reduces, which affects its price.

- Water becomes hard due to pollution.
- Wastes from the acid factory thrown into the near-by canal burns crops in the near-by field. This also affects fertility of the soil.

Sagarpur
- Length of wheat crops has shortened. There is also less number of grains per wheat crop, which are relatively thin. This is because dust from brick kiln falls on the crop.

Uncha Gaon
- Black deposit (assumed to come from local thermocol factory) falls on cauliflower and spinach, which is more prominent during winter. This leads to the cauliflower turning creamish and the spinach turning blackish. The taste becomes salty. Such loss is estimated to be 10 per cent for cauliflower and 5 per cent for spinach.
- Tomatoes were also turning blackish though not resulting in financial loss.
- Flowers of brinjal, tomato and chili get wasted when the ashes along with smoke fall on them.
- As per men group, Moriya disease in chili is assumed to be due to some pollution in the air.
- The water also turns black when kept overnight.

- The men group referred smoke as carbon.

Baroli
- The stem of jowar and wheat leaves when dried as fodder turns black,

which as per men group should be white in colour.

- Men group perceives that increased growth in pest and wheat is due to weather fluctuations, which is due to air pollution.

Pali Kasba
- Length of wheat crop has shortened. With dust and fume from nearby factories falling on the crop, it damages the crop, which slowly starts decaying.

Sohtai
- There is black powder deposit on the leaves at Sohtai.
- Land is affected through pollution from factories at Chandawali.

- Air pollution is mainly from the two brick furnace that exists in the village. Such factories operate in all seasons excepting the rainy season.

Sahapur Kalan
- Environment unfriendly wheat is on the increase.
- Crop disease is spreading faster than before.

- There are no local factory in this village, however, there is polluted air from Chandawali factories.

Chandawali
- In winter, impact of smoke and dust is high.
- Crops like mustard and cauliflower are more vulnerable.
- A thin layer of smoke is observed on the leaves, more in winter season.

- Two groups of men and women pointed out that maximum pollution was from brick and cement factory.
- Women said that while cleaning roof they find thin layer of black deposits.

Sumper
- Polluted air blowing from North-South direction (Ballavgarh and Malerna) damage crops and vegetables like mustard, coriander and cauliflower.
- A thin layer of smoke is visible in the field even on a clear day; the smoke is prominent in winter and does not allow sun rays to reach the crop and makes pesticide ineffective.

Piyala	• In the Western side of the village, around 1 kilometer away, there are few clusters of brick kilns which is creating a lot of dust pollution in nearby agricultural fields. Dust and smoke from brick kilns get deposited on the aerial parts of vegetation and also on land.	• Such dust and smoke hinders growth of crop and affects yield. The farmers found a difference in yield of wheat to the extent of 3 quintals per acre between fields near the brick kiln and those located away from such kilns.
Jhajru	• There are a few factories in the adjoining areas (not that close to the village), which give rise to obnoxious smell. An unauthorised and unknown drug/chemical factory on the G.T. Road set up since 3 to 4 years, at distance of one kilometer emits gas pollution. Air pollution is also due to nearby railway line and dust pollution is due to soil being dug up for sale.	• Ripening and filling of grains are affected due to polluted air and yield is also affected. Since there are not many factories in and around Jhajru, the village groups were not able to provide much correlation between pollution and its adverse impact on agriculture.

Note: As described by women groups and men groups from selected villages of Faridabad

Source: Based on field reports on Faridabad by PA Researchers

3.6 Health Status at Faridabad

Community perspective on general health status at Faridabad villages indicates the prevalence of impact of air pollution apart from other common diseases. The community members are aware of their deteriorating health condition in recent years rising with rapid industrialisation at the local level. This has been listed in table 3.5. There has been a rise in respiratory disease, breathlessness, burning sensation of throat, eyes and nose and headache, with higher incidence in winter months, when the level of pollution increases. It is not difficult for many village groups to draw one to one correspondence between local emissions and physical discomfort and illness, which they are forced to undergo. The health expenses have gone up for local communities since home medicines hardly work. Health of their livestock is also affected by air pollution causing indigestion, diarrhea and death.

Table 3.5   Listing of Health Issues, Faridabad

Name of Village	Type of Disease/Illness identified
Sohtai	• Women group/s reported increase in respiratory problems. • Illness in the area has increased in the last 5 years, which the community attributed to use of improved variety of seeds and fertiliser and also to the brick furnace and other factories around the village.
Kadhaoli	• Burning sensation in throat, eyes and nose; pollution causes throat irritation and cough and when ignored results in TB; weakening of eye sight due to fumes from near-by pharmaceutical factory; giddy feeling due to fumes emitted from rubber and pharmaceutical factory. • Mysterious death of buffaloes during the last two years in monsoon is attributed to air pollution.
Sagarpur	• Eyesight getting affected amongst others by brick kiln emissions; rise in cases of TB and cancer; lethargy, greying of hair, breathing problems and fever especially among children. • Mysterious death of buffaloes during the last two years in monsoon is attributed to air pollution.
Uncha Gaon	• There has been increased incidence of burning sensation in the eyes and TB for the last 10-15 years. Air pollution, according to men group, has led to increased incidence of heart attack and cancer. Women attribute increased incidence of breathing problem and coughing due to air pollution. Emissions from rubber and thermocol factory lead to bad odour and feeling of nausea.

	• Men group felt that the pollution from thermocol factories might be affecting the health of their buffaloes though no visible symptoms have been noticed. The women, however, do not think in similar fashion.
Baroli	• Increased incidence of "Najla", breathlessness and burning sensation in eyes. There also takes place fluctuations in blood pressure. 50 per cent of the population suffers from "Najla". Breathlessness is high during the months of September, October, November, March and April.
Pali Kasba	• The local community members have incidence of skin disease, TB, cancer, heart attack, indigestion, diarrhea, nose burning, headache and dizziness.
	• Air pollution causes indigestion and diarrhea in livestock and buffaloes refuse to eat fodder made of crops grown in the field.
Chandawali	• Illness is on the rise, home remedies do not work and people have to depend on doctors.
	• Complaint of TB has become common since last ten years.
Sahupura	• Men group/s listed some common ailments, which have increased over the last 7-8 years. These are blood pressure, heart ache, TB (persistent coughing results in it), "baye" (arthritis??), pain in appendix (caused by driving tractor), sudden death of youth at young age (30-32 years), allergy(from medicines and congress grass-more in Kuar and chet), cancer, weakening of eye sight, asthma, cold and cough.
	• Women group/s listed common ailments affecting women. These are malaria (its incidence has fallen over the years), dengue (incidence for the first time two years back), acidity (due to poor quality food, more after child birth and sterilization operation); "baye" (arthritis, more in winter and monsoon), head ache and pain in stomach, young females have serious cases of leukorria (white discharge), pain in the waist, tooth decay among children (pest such as green caterpillars in the crop cause it), "najla" (cold, cough,

	running nose, headache—more during monsoon and winter months), breathing problems (due to polluted air from near-by Ballabgarh industrial estate.
Jhajru	• Local pollution of gas causes irritation in breathing and uncomfortable feeling. High levels of dust pollution cause irritation in nasal tract and eyes, especially during winter months.

Note: As described by women groups and men groups from selected villages of Faridabad. The women and men groups associate increased incidence of health problems due to pollution. Some of them also point out the poor quality of cereals, pulses, vegetables, fruits and dairy products due to increased use of fertilisers, which causes lower nutritional levels and leads to health problems. When asked whether rising health problems were due to increase in awareness of health, some of the participants agreed but also remarked that there was sharp increase in incidence of illness and diseases. Increase in incidence of cancer and TB was correlated with factory pollution apart from consumption of fertiliser-based crops

Source: Based on field reports on Faridabad by PA Researchers

Faridabad—How Effective is the Support System?

At present the existing support system in Faridabad is not quite responsive. There are few community-based NGOs working in the villages covered under research and their mandate is limited. For instance, in Sahupura village, a people's organization called Jan Kalyan Sewa Samiti was registered in 1992 by village youth, which looks after village cleanliness but not agriculture issues. However, during discussions held by village groups with PA researcher, the organisation got interested in pursuing issues related to the agriculture department. A similar organization exists in Sahapur Kalan, which is not involved in agriculture-related activities.

For many villages, local level institutions like local Panchayat is not that effective for lack of power of decision making and funds. Land-related conflicts are very common and people take help of the local magistrate's office or the police. The sub-divisional agriculture office at Ballabgarh caters to 192 villages. Visits of extension officials are not regular and do not help in solving local problems. Visits of officials from the department are few, which benefit generally the well to do farmers while others remain unaware of such visits. Indifferent attitude of the local state departments related to agriculture is a major stumbling block. Irregular supply of fertilisers and pesticides results in crop loss. Credit agencies, like bank and local co-operative also do not respond much to farmers' needs. Such problems are more for the small and marginal farmers. Two illustrations from Unchagaon, Faridabad and Baroli, Faridabad are given in boxes 3.8 and 3.9.

Box 3.8
Village Unchagaon: Institutions not Helpful

The men group of village Unchagaon, Faridabad, informed that the Sarpanch or the village head is not at all helpful and was least bothered about their problems. The villagers are indifferent to the existence of government agriculture office because they do not have the time to visit the office and gather information. The officials rarely visit them. Rather they visit only the rich landlords. At the annual fare organized by the agricultural office invitations are sent to a maximum of 20 cultivators and the same cultivators are invited every year. However, the office is helpful when some information is sought from them and it also endorses their forms for procuring fertiliser from the cooperative shops. The BDO plays a passive role and is only approached at the time of crisis. Wholesale market is the place where they sell their produce. The relationship with the private dealer is important for he offers loans at the time of crisis. Private dealers in addition to lending also play a crucial role in providing information on the latest in the market and also on ways of cultivation.

Source: From Field Notes on Faridabad villages, Sudipta Ray

In one Venn diagram drawn by a village group of well to do farmers at village Sahupura, provides interesting results. In the Venn diagram, the social relationship of the village with other agencies is indicated. According to the village group the perceived relationships is as follows.

- Market is the most important place for them as they procure all their requirements like seeds, fertiliser and pesticides from the market. The market is within their access.

- Ballabgarh grain market is another place where they sell their products. It is almost in equal importance as compared to the market from where they procure seeds and fertilisers.
- Office of the Sub-Divisional Magistrate is another place where they go for settling their land disputes. But it is relatively less important and is at a greater distance.
- Village Panchayat is of very little importance as it is not capable of rendering much help.
- Rural bank is perceived as important and is at a relatively closer distance.
- Electricity department is very important for getting their land irrigated but it is beyond their reach. Hence, it is shown at a greater distance.
- Cooperative society that provides seeds and fertilisers is the most important organization for them.
- Mortgage bank provides loan on land and gold for land cultivation and procurement of agricultural implements.

In the process of Venn diagramming, as described above, the perceived social relationship of villagers with different institutions is shown. Cooperative Society is perceived to be most important and village Panchayat the least important institution. Electricity board is considered to be the most distant while seed and fertiliser market and bank are perceived to be the closest.

Box 3.9
Village Baroli: Institutions Oppressive

The farmers of Baroli village are concerned about their development of agricultural situation but they do not find immediate solution. Haryana Urban Development Authority (HUDA) is proposing to take over their lands. For the farmers from Baroli air pollution is less important than problems like pest invasion and fragmentation of land. People are critical of panchayat for the money and power provided to panchayat by the government for development of village are not used effectively. For example sarpanch of village Baroli had deducted Rs.100/- from widow pension for construction of toilet but the people have no idea as to when the work will start. There is a voluntary association of youth, which helps panchayat in different social work however this is not enough. The local farmers suggest that that there should be more transparency in panchayat activities and they should be informed about such activities and utilisation of funds from time to time. About agricultural extension programme, it should be introduced only after assessing farmers' needs through meetings in gram sabha. All fake seeds, fertilisers and medicines should be removed from the market and the prices of agricultural inputs should be controlled.

Source: From Madhumita's report on Baroli village, Faridabad

Residents from different villages have taken various steps to address agricultural constraints. In Malerna, they sent communication to the Panchayat to take action against the local thermocol factory. However, no action has yet been taken apparently because the factory owner was an influential person. In village Chandawali, villagers complained against the owner of the brick kiln, but the Panchayat could not take any action as the owner had a legitimate certificate for the factory. Villagers of Sahupura and Malerna went to the Commissioner of the district who promised to act on their complaint against HUDA's acquisition of cultivable land at low rates but no action was communicated to the villagers.

Cooperative banks, other banks and sub-division agriculture office at Ballabgarh are a few institutions that are meant to support agriculture activities. Often seeds and fertilisers are not available on time from the cooperative bank and the farmers are forced to depend on private shops and procure adulterated items at high price. Banks give credit but require elaborate formalities to be completed by farmers before procuring loans. Agriculture Development Officials from Ballabgarh occasionally visit the village to get in touch with the farmers, though few have prior information.

As regards air pollution, there have been a series of efforts undertaken at different levels for controlling such pollution in and around Faridabad industrial town. Faridabad Industries Association (FIA) is actively involved in greening the city. There are strict norms and conditions specified by FIA for industries to maintain vegetative cover in and around their factories. The State Pollution control board also monitors the level of toxic elements in the effluents of the factories. However, air and noise pollution are far in excess of permissible limits in the urban areas of Faridabad and also extends to the rural areas.

3.7 Impact of Urbanisation and Industrialisation on Quality of Life at Faridabad

In many hamlets, majority of people felt that the process of local industrialisation has helped them by offering more employment, a larger market and higher prices for their products such as milk and food grains. Although the costs of industrialisation has been considerable as shown in table 3.7. Such costs relate to scarcity of grazing land, loss of village identity, food insecurity, livelihood problems, health complications due to pollution, increased cost of living, congestion, social problems, damage to crops etc.

Box 3.10
Oral Assessment of Urbanisation and Industrialisation by Ram Kaur,
a Dalit woman agricultural labourer from Khandawali, Faridabad

- Agricultural activity is fast declining because of proliferation of factories and industrial units. There is chemical pollution and smoke emitted from acid, which causes health problems like cough, TB etc. In the night, there is gas/smoke emitted from plastic factory, which makes our eyes red. After every 15 days some factories emit poisonous gases, which make for swelling of eyes. In the last 4/5 months back, the eyes started burning, eyes swelled. The carbon emitted from the boilers gets deposited on our rooftops.
- A powerhouse has been constructed nearby with capacity for high power voltage. It can draw people from 5 ft. All wires pass through the village. If there is some accident at the powerhouse then it can cause extensive damage to people living here.
- Water is hard from the factory. Earlier, the food used to be good and tasty. Now food has fallen in value and nutrition level.
- We do not keep livestock because we cannot arrange for fodder. Earlier share cropping was possible but now due to mechanisation the

---

chances have got reduced. The local factories prefer wage labourers from Bihar and Nepal not from Haryana at monthly wage of Rs.1000/ Rs.1200/Rs.1500/—whereas Haryana labour wants Rs. 2500/-. So we have experienced no benefit from local industrialisation.

Source: Field Notes on Faridabad villages, Neela Mukherjee

Table 3.6   Perceived Costs/Benefits of Urbanisation and Industrialisation, Faridabad

Names of Villages	Benefits	Costs
Malerna	• None listed.	• With HUDA constructing houses next to the village everything will become expensive. There will be no place left for grazing of livestock and the village identity will be lost.
		• Resisting acquisition of land by HUDA is ineffective. The concern is that such acquisition of land would lead to scarcity of food at the household level and affect the survival of livestock.
Kadhaoli	• With increased demand for land for non-agricultural purposes, price of land is rising.	• Agricultural land has been acquired leading to livelihood problems.
		• Air and water have got polluted.
Sagarpur	• Industrial area has provided employment opportunities and higher income.	• Few members of local community perceived increased pressure on them

- As a spillover effect development activity has also taken place in the village.
- Growth of urban centre has led to increase in knowledge about urban facilities.
- Selling of milk has become profitable and scope has been created for opening of units to produce dairy products.
- Demand for food, vegetables has increased.

(those not employed in the industrial/urban sector) to earn more in order to meet increased cost of living and keep parity with rising standard of living.

Uncha Gaon

- Growth rate of development is faster with employment opportunities getting created.
- Better facilities are available with nearness to urban centre.
- More information is now available about national and international affairs.
- With transport facilities and railway line, marketing of crops and vegetables is easier.

- Due to expansion of industrial/urban area, the land of the local community has been acquired.
- Now industries do not provide them with enough employment as earlier. Now the employment is on contract basis with wages below minimum.
- Due to urbanisation, expenditure on clothes, cosmetics, chocolates and toys have increased.
- With rise in cost of living due to industrialisation, many local community members are forced to sell milk rather than feed it to their children.
- The senior members of the community perceived that due to urbanisation joint family system are breaking down.

		• Shortage of water and electricity was perceived as adverse impacts of industrialisation; amongst the other impacts, congestion was another problem and deterioration in conditions of roads and surroundings was still another.
Baroli	• Employment opportunities created through urbanisation. • Younger generation has now the opportunity to learn new technologies. • Good price received for agricultural produce and milk.	• None listed.
Pali Kasba	• Rise in income and employment for the village is used as a transit point from Ballavgarh to Delhi and some nearby mining areas. • Profitable to sell milk in the Faridabad town.	• The lands of the farmers have been acquired thus affecting their livelihood. • With use of water for mining in near-by areas, water level is going down thus adversely affecting availability of water. • Health complications due to industrialisation are leading to loss of working days.
Chandawali	• Vegetable cultivation is on the increase because of its market demand and it earns ready cash.	
Sahupura	• Proximity to the industrial estate and urban centre will reduce transportation cost and will also provide ready market for vegetables in the future. • Non-agricultural work mainly labourer's job in Ballavgarh and	• The men group pointed out that air and water pollution from the industrial estate acts as a constraint. Air pollution has damaged mustard flowers over the last 5 to 6 years forcing

Faridabad town absorbs youth of the village in absence of any permanent employment.

reduced cultivation. The pollution can cause loss of agriculture produce to the extent of 75 per cent.
• Water pollution of canal water burns the crops growing in nearby fields.

Note: As described by women groups and men groups from selected villages of Faridabad
Source: Based on field reports on Faridabad by PA Researchers

Box 3.11 provides an outline of future quality of life and living of the women of Chandawali as neatly summarized by them.

Box 3.11
Fear of Losing 'our' Wealth by Urban Expansion,
Chandawali Village, Faridabad

Women group doing agriculture in Chandawali village under urban expansion in Faridabad district of Haryana, feel powerless against the rapid pace of industrialisation. They emphasise that social costs of urbanisation and industrialisation outweigh the benefits and have visually illustrated their thoughts about the bleak future for themselves and their children. They drew three pictures (Chart 5a, 5b and 5c) to illustrate the point to show how their village Chandawali was 40 years back, how it is now and what will it look like in the future, after 40 years.

Earlier, 40 years back they owned land, bullocks to plough, good crops, trees, water, well, school for children and their skills in artisanship. Today, they have little or no land. Their agricultural land is being gradually taken up for industrial units, like cement factories, brick kilns etc. They work as labourers in such factories while outsiders/the local urban department of the government is acquiring land at low price for construction of houses. After 40 years, there would be multi-storey flats in the locality built by the urban department of the government and sold to outsiders. The original residents would lose all their land, their fields and their livestock and would be forced to work as household maids/etc. in the homes of flat owners. There would be a nice big park and super market to cater to the 'well to do' residents living in those flats. But the erstwhile villagers would have nothing to look forward. They and their children would lose their identity, struggle for survival, hard life and live in near-by slums and work as maids/labourers in the multi storey flats.

Source: Village Report on Chandawali by Bratindi Jena

3.8 The "Do-Able's" at Faridabad

Local community members of the villages under study at Faridabad suggested the following "do-ables" for overcoming the problems and constraints faced by them and also in improving their quality of life. The "do-ables" are those, which the local village communities expect the government to do. It has little or no relationship with what the community members can do themselves. The emphasis is more on the expected role of the government and its concerning agencies on three kinds of generic issues as follows.

- Offsetting Damages from Industrialisation and Urbanisation.
- Enhancing Agriculture Support System.
- Strengthening Agricultural Policy Support.

Offsetting Damages from Industrialisation and Urbanisation

- Construction of houses on agricultural land by HUDA should be discouraged/stopped.
- Displaced farmers and other members of local community must be provided with alternative employment rather than cash compensation.
- Those local factories, of the polluting type, should not be given permission to be located near agricultural field and/or within the village.
- Public roads should not be constructed near crop field as it leaves dust deposit on leaves along with pollution.

Enhancing Agriculture Support System

- The agricultural support system needs to be strengthened and its services ensured at village level. Extension officials should meet more farmers in the villages rather than a few. Timely supply of seed, fertiliser and rat poison should be ensured.
- Proper facilities should be provided in terms of regular supply of electricity, water and other inputs to encourage land cultivation. Irrigation facility needs to be increased by constructing six sub-canals from Agra canal for irrigation purposes.

Strengthening Agricultural Policy Support

- Prices of agricultural inputs like fertiliser, pesticide and seed can be controlled by the State.
- Local (old) variety of crop seed should be promoted.
- Farmers should have a right to decide the price of their own agricultural product.
- Protected animals like nilgai (blue bull) should be kept in sanctuary.

Annex 'B' Profile of Villages at Faridabad

For participatory field research, the villages of Faridabad were chosen on the basis of areas falling in the zone of air pollution as given by scientific data. However, the villages have wide variations in terms of factors such as size, socio-economic, cultural, ethnic, agricultural, institutional, locational, environmental and pollution-related factors. Some villages such as Baroli and Pali Kasba are interior villages while Khadaoli is a roadside village. In terms of size Malerna has 350 households while village PaliKasba has 122 households. The caste/community composition of the villages is quite different though most villages have multi-castes residing there. Both Sagarpur and Piyala villages are Jat-dominant in terms of caste while Khadaoli is muslim-dominant and Pali Kasba is a Gujjar-dominant village. In some villages like Khadaoli, agriculture is the main source of livelihood as compared to village Jharsainthly where agriculture is a subsidiary source of livelihood. Though the generic issues in agriculture are common to most villages, the priority issues are not always the same. The impact of air pollution is perceived strongly in some villages as compared to the others. Though all villages fall in the air pollution zone, their make up in terms of socio-ethnic, geographical, agricultural set up and source/s and impact/s of pollution vary widely.

Short profiles of urban and peri-urban villages covered during field research are given below. Such village profiles are mainly centered around perspectives of respective village groups on socio-economic aspects, agriculture and livelihood and issues in agriculture and air pollution.

Name of Village	—	Baroli
Name of Block	—	Faridabad
Name of District	—	Faridabad

Socio-Economic Profile The village is located approximately 1.6 kilometers southeast of Faridabad Industrial Area and 2.6 kilometers of Ballabgarh Industrial Area. It is an interior village not on roadside. Its total population is 6000 and has a total number of 600 households. There are 5 wards in the village and the dominant caste is that of Gujjars. The Gujjars have 440 households, the Pandit has 40 households while Dalit and tribal have 120 households. The village has 5 Angadwadis, 1 veterinary hospital and 1 government-run senior secondary school.

Agriculture and Livelihood The main livelihood sources in the village are agriculture, milk and private service. The village has a total cultivable area of 440 acres. Its soil is clay. In Baroli, Kharif crops grown are jowar, bajra, maize and dhencha and in Rabi, crops grown are wheat, mustard and barsam.

Issues in Agriculture Some issues in agriculture relate to short supply of electricity resulting in poor irrigation of jowar and wheat, damage to crops by locust, monkey, nilgai, cows and termite and increased attack of pests and weeds on standing crops.

Issues Related to Air Pollution Air pollution from the Haryana Thermal Power Station is related to Najla, breathlessness, burning sensation in eyes, fluctuation in blood pressure. No impact has been observed on crops except blackening of stem of wheat and jowar. Air pollution affects weather, which thereby impacts on pest and weed and makes them grow faster.

Name of Village	—	Chandawali
Name of Block	—	Ballabgarh
Name of District	—	Faridabad

Socio-Economic Profile Village Chandawali is situated 6 kilometers east-west direction of Indian Oil Corporation at Faridabad and 3 kilometers away from Ballavgarh. It is a multi-caste village with 750 households belonging to Jat, Gujjar, Brahmin, Dhobi (washerman), Manihar (bangle maker), Potter, harijan, Kohli and Saini. There are 11 wards in the village and people are settled in 8 caste-based clusters where Jats are in two clusters, Gujjars are in one cluster and all other castes are in individual clusters.

Agriculture and Livelihood The total cultivable area in the village is 800 acres. There are 45 tractors and 150 tube wells in the village. In this village, soil is fertile and good for vegetable cultivation. Some crops cultivated in this village are wheat, potato and paddy. Vegetable cultivation is on the increase because of increased market demand and availability of ready cash. The residents of village Chandawali are dependent on both agricultural and non-agricultural activities, almost in equal proportion of 50 per cent. Jat, Gujjar and Yadav communities are the main land-owners. People from other castes are engaged as agricultural labourers, share croppers (about 5 per cent) and work in nearby towns. Income-wise, 60 per cent comes from agriculture and livestock and 40 per cent from jobs and other occupations. Local factories provide less employment scope. Land acquisition by HUDA is depriving people from land cultivation.

Issues in Agriculture The main agricultural constraints in the village are irregular supply of electricity affecting irrigation, pest attack on crops, plant disease, non-availability of fertiliser on time and blue bull attack. Incidence of pest and disease has increased with introduction of new variety of seeds. Blight disease is common in the potato crops such as Kufri 5857 and Kufri Anand. To control such disease, medicines like Roger and M-45 Melathin are applied.

Issues Related to Air Pollution Two groups of women and men interacted with have observed pollution from the brick and cement pipe factory. Due to increased demand for bricks the capacity of the brick kiln has been doubled. The women group reported that while cleaning their roofs they found black water. A thin layer of smoke was also observed on leaves. Such pollution impact is more in the winter season. Large quantities of vegetables cultivated in the village get affected due to air pollution. With regard to health incidence of TB has become common since last ten years.

The village groups suggested that acquisition of cultivable land for construction purposes should be stopped and alternative employment should be offered to displaced person instead of cash. The village groups also stated that setting of factories in the locality should be banned since they have adverse impact on human beings.

Name of Village	—	Jhajru
Name of Block	—	Ballabgarh
Name of District	—	Faridabad

Socio-Economic Profile Jhajru village is located 12 kilometers from Faridabad on the Grand Trunk Road and 3 kilometers in the north east of highway. The village has high literacy rate of 80 per cent of both women and men as compared to other villages situated in its vicinity. It has a population of around 3000 to 3500 in 300 households. The total geographical area of the village is 500 acres. There are different castes in the village with 150 households of Jats, 10 households of Brahmins, 70 households of Dalit, 20 households of Carpenter, 20 households of Kumhar (Pottery), 10 households of tailors and 20 households of Banias (traders). There are 4 mohallas in the village, Loupa Patti, Bis Patti, Nichala thok and Harijan (Dalit) Mohalla. There are two middle schools in the village, one is a government school while the other is a private and there is also an Anganwadi.

Agriculture and Livelihood Agriculture forms the major source of village income, followed by services. Such agricultural income accrues to big and medium farmers, while small and marginal farmers produce for self-consumption and get income from animal rearing and doing petty jobs. In occupational pattern of the village, agriculture accounts for 50 per cent, labourer 30 per cent, service 15 per cent and business 5 per cent. There is general mono cropping and wheat is the major crop to be cultivated. Some households grow dhaincha as green manure and jowar as fodder.

Issues in Agriculture Some of the major issues in agriculture relate to alkalinity of underground water, which prevents mixed cropping and crop rotation. This is in addition to erratic electricity supply and uncertainty in timely availability of fertilisers, which adversely affects crop growth and yield.

Issues Related to Air Pollution Increased spate of industrialisation affects ripening and filling of grains and loss of yield. However, people were unable to offer any concrete reason since a few factories can be seen in and around Jhajru. When the wind blows in the direction of the village there is the problem of bad odour. It was reported that one unknown and un-authorised drug/chemical factory on the G.T. Road at one kilometer distance has been functioning on the Grand Trunk road. The gas emitted from this factory causes some irritation in breathing and leads to discomfort. Air pollution is also caused by the near by railway line and the dust particles due to soil being dug for sale. The inhalation of dust causes irritation of nasal tract and eyes, mostly felt during winter season.

The farmers think that the problems related to agriculture need urgent action. They expect the government to provide basic infrastructure facilities for the development of agriculture and the village. Social problems need to be tackled at the village level especially at the level of village leaders though there is not much unity amongst the villagers.

Name of Village	—	Jharsainthly
Name of Block	—	Ballabgarh
Name of District	—	Faridabad

Socio-Economic Profile Village Jharsainthly is situated on the main Mathura Road, nearly 11 kilometers from the main Faridabad town and 2 kilometers from Ballabgarh. The village is on both sides of the Grand Trunk road and surrounded by a cluster of industries, both big and small. Jharsainthly has an approximate population of 6000 with 800 acres of geographical area. It has a number of castes residing in the village such as Jat (300 households), Kohli (50 households), Brahmin (10 to 12 households), Barber (Nai) (17 to 18 households), Sweeper (20 households) and Potter (Kumhar) 10 to 12 households. The village has a middle school, three branches of nationalised banks, piped supply of drinking water and electricity. The literacy rate in the village is 30 per cent for age group below 40.

Agriculture and Livelihood Jharsainthly is a part of Faridabad Municipal Corporation. HUDA has acquired much of the village land, especially agricultural land. The remaining cultivable land is in patches, away from the main village. The crops grown in the village are wheat, jowar, vegetables and bhindi. Around 70 per cent of village land is acquired by HUDA, with only 20 per cent land left for cultivation. This land is affected by alkalinity and wheat crop, which can stand alkalinity, is grown during Rabi season. The land is kept fallow in zaid season, while its utilization in kharif depends on rainfall pattern.

Issues in Agriculture Agriculture is a subsidiary source of income in the village and such land is slowly being diverted for non-agricultural uses. Some agricultural constraints are shortage of agricultural labour (especially for big farmers) and input supply, adulteration of fertilisers and pesticides, poor quality seeds, pollution and some irrigation problems (for small farmers). The farmers said that during the past season brinjal and tomato crops were affected by fruit borer. They also said that though different variety of seeds was available the seed traders were not ready to make any guarantees about germination and quality of products.

Issues Related to Air Pollution A few farmers pointed out that pollution problem is limiting agricultural production, especially those farmers having their field near sources of vehicular and industrial pollution. Mustard crop is completely eliminated from the area on account of dust and smoke, which prevented the crop from setting seeds. Farmers also reported that the grain size of wheat crop is reduced and now they are getting shriveled grains. Vegetable plants like cauliflowers are affected by air pollution and turn blackish and do not fetch good market price.

Name of Village	—	Kadhaoli
Name of Block	—	Ballabgarh
Name of District	—	Faridabad

Socio-Economic Profile The village is located approximately 6.5 kilometers west of Faridabad Industrial Area and 4.3 kilometers west of Ballabgarh Industrial Area. It is 2 kilometers west of Ballabgarh new industrial estate. It is a roadside village. The village has a total area of 172 acres. Its total population is 3000 and has a total number of 400 households. There are 2 wards in the village and the dominant community is that of Muslims. There is also a cluster of Dalit residing in the village. The village has 1 Angadwadi and 1 Government-run primary school. The village has 4 factories—a plastic factory manufacturing plastic bags, two rubber processing factories and a cement brick kiln factory.

Agriculture and Livelihood The main livelihood sources in the village are agriculture and milk. The village has a total cultivable area of 114 acres. In

Kadhaoli, Kharif crops grown are jowar, maize, arhar, paddy, moong, drencha and sesamum, while in Rabi crops grown are wheat, mustard and vegetables like cauliflower and radish.

Issues in Agriculture Some issues in agriculture relates to short supply of electricity, locust and termite attack, damage of crops by nilgai, increase of weeds and insects and hardening of water.

Issues Related to Air Pollution Besides the polluting factories in the village, the village groups also considered factories in the vicinity as emitting pollution. The polluting factories nearby the village were pharmaceutical factory, thermocol factory, scrap iron processing factory, iron rolls factory and textile factory. The fumes from the polluting factory get deposited on wheat, especially in January and February. Water pollution leads to hardening of water, which affects the quality of soil. There is burning sensation in throat, eyes and nose. There is weakening of eye sight due to fumes from pharmaceutical factory and feeling of giddiness from fumes emitted from rubber factory and others.

Name of Village	—	Malerna
Name of Block	—	Ballabgarh
Name of District	—	Faridabad

Socio-economic Profile Village Malerna is 5 kilometer away from Ballabgarh and 4 kilometers away from the Faridabad scientic site of the project. It has a total adult population of 1768 women and men with 350 households. The village has four clusters named after the caste that reside. Yadav and Brahmin reside in one cluster, Dalits (shoe maker, scavenger and potter) reside in another while Jath and Kander (weaver) live in separate clusters. Yadav, Brahmin and Jath are agriculturists. There are two 'anganwadis' in the village, one for Dalits and the other for weavers. There is also a primary school.

Agriculture and Livelihood In this village, livelihood is dependent on 50 per cent on farming and livestock rearing and 50 per cent on labour. About 30 per cent of labour are employed in the near-by factories, 10 per cent are on daily wages and 10 per cent migrate on a seasonal basis. The village has a total cultivable land of 500 acres with 40 tube wells and 20 tractors. About 15 per cent of land is cultivated through share-cropping. The water is hard and the soil is do moth. Wheat is the main crop of the village apart from millet and paddy, which were introduced five years back. Vegetable is cultivated in small quantity.

Issues in Agriculture As stated by village groups in Malerna, problems and constraints in agriculture include irregular electricity supply, which disrupts irrigation facility. Fertilisers and pesticides are not available in time and adulterated ones are not that effective. Parthenium weed has spread across the fields in the last three years thus affecting crop production. The new variety of crop is more vulnerable to pests and disease. Stray cows and nilgai (blue bull) also cause loss in harvest. HUDA's acquisition of land in the area is the biggest constraint to agriculture. With HUDA houses in the vicinity living will become expensive. It is a threat to village identity. It will lead to scarcity of food and affect livestock rearing. There will be no place for grazing livestock.

Issues Related to Air Pollution Factory smoke is perceived by village groups, of which, one harmful smoke is emitted when factory waste is burnt fortnightly, on the roadside. There are smoke deposits on the leaves, which when consumed by livestock also affects the smell of milk. The noise pollution through machines/generators run by factory is also high and complaints to factory owners and administration have not had any consequence. The impact of such pollution is also felt on health where asthma is a common complaint. Some people are also getting TB. Since the last ten years the no. of diseases have increased. In the past there were less vehicles and no factories in the village. The factories were established without permission of local people and now there are no forums to listen to their complaint.

The major suggestions by village groups from Malerna are that no factory should be set up in the vicinity of the village; HUDA's activity of acquiring agricultural land for house construction should be stopped; electricity supply should be un-interrupted for irrigation; un-adulterated fertiliser and pesticide should be made available on time; and nilgai should be protected in sanctuary.

Name of Village   —   Pali Kasba
Name of Block   —   Faridabad
Name of District   —   Faridabad

Socio-Economic Profile The village is located approximately 7 kilometers north west of Faridabad Industrial Area and 8 kilometers of Ballabgarh Industrial Area. It is an interior village, not on roadside. It is a large village with a total population is 7200 and has a total number of 1200 households. There are 3 wards in the village and the dominant caste is that of Gujjars. The village has 3 Angadwadis, 2 Government-run senior secondary schools, 1 Industrial Training Institute, 1 Agriculture Support Office and 1 Women's Sewing Training Centre.

Agriculture and Livelihood The main livelihood sources in the village are agriculture and milk. Others include agricultural labourers, business, government service, private service holders and truck operators. The village has a total cultivable area of 3000 bigha. In Pali Kasba, Kharif crops grown are jowar, bajra, maize and dhencha, while Rabi crops grown are wheat, mustard and barsam, safed jai and gram. Vegetables are also grown like cauliflower, cabbage, spinach, radish, potatoes, methi and onion.

Issues in Agriculture The major issues in agriculture relate to short supply of electricity, crop damage by nilgai, rats, termites and stray cows, vulnerability of crops to frost, hailstorm and irregular rainfall, locust attack and attacks of weeds and pests.

Issues Related to Air Pollution Length of wheat crop has diminished in recent years. The men group opined that this is due to dust and fume falling on the crops, which leads to crop wastage. Such dust and fume are sourced from mines and near-by factories. The fumes from coal tar mixture plant, JAV steel recycling and plastic recycling plant and dust from stone-crushing zone is perceived to be harmful by the villagers. TB is on the rise for the last 10 years. Headache, burning sensation in the nose and feeling of dizziness and cancer cases are on the rise for the last 4 years. Increased incidence of itching, sores and swelling amongst the residents is there for the last 4-5 years. Crops from the field, when consumed, is increasingly resulting in diarrhea and indigestion for the last two years.

Name of Village	—	Piyala
Name of Block	—	Ballabgarh
Name of District	—	Faridabad

Socio-Economic Profile Piyala village is around 18 kilometers from Faridabad Industrial town, of which 15 kilometers is on Delhi-Mathura road and 3 kilometers is relatively interior on the North-East side of Delhi-Mathura road. The village is connected by pucca-metallic road. A part of the village land is under public sector industry such as BPCL (Bharat Petroleum Corporation Limited) and private industrial houses like Balmer Lawrie Limited. The population of the village is around 5000 with households numbering 400. Caste-wise distribution of households is that of Thakurs in 80 households, Jats in 1250 households, Brahmins in 40 households, Muslims in 25 households, Lohar (blacksmith) in 5 households, Nai (Barber) in 5 households, Balit, Kohli and Chamar in 70 households and Bhangi in 40 households. The village has a high school and a primary health centre.

Agriculture and Livelihood One main source of livelihood in Piyala village is that of providing service in which around 30 to 35 per cent are involved. Agriculture constitutes around 25 to 30 per cent of livelihood and daily labour makes for 30 to 32 per cent. The other sources of income include livestock rearing, weaving, tailoring, vegetable trading etc. The main crop of the village is wheat in Rabi season, which provides work in sowing and harvesting activities. Though not sufficient, some land is irrigated through canal water and some through tube well. Around 70 per cent of total village area is utilised for agricultural purposes, though seasonally. Since agriculture is mainly at subsistence level, marketable surplus is low. Most of the produce from agriculture is for self-consumption. With provision of drinking water facilities, vegetable cultivation is also being undertaken in home gardens. Sharecropping and contract cropping is also prevalent.

Issues in Agriculture Some major issues in agriculture are related to inadequate irrigation and alkalinity of underground water. The problem of Jai weed in wheat is almost intractable. Uncertainty in timely supply of quality inputs like fertilisers and pesticides increases the risk of cultivation.

Availability of labour is generally a constraint for big farmers. The low compensation rate of the government for acquiring land is also a problem. There is also absence of proper marketing linkages for sale of marketable surplus from the village. The problems are acute for the poor farmers.

Issues Related to Air Pollution Piyala village has clusters of brick kilns in the western side of the village, which creates considerable dust and smoke pollution in agricultural fields. Dust and smoke from brick kilns get deposited on the aerial parts of the vegetation and sometimes on the land. It arrests crop growth and has an impact on yield. Experience shows, there is a difference of 3 quintals of yield per acre between fields near brick kiln and those situated away from it. In Piyala village sometimes there is acute air pollution especially when wind is blowing from east to west towards the village. The smell is of leakage/processing of gas in gas filling plant and it leads to nausea and headache of local people.

Name of Village	—	Sagarpur
Name of Block	—	Ballabgarh
Name of District	—	Faridabad

Socio-Economic Profile The village is located approximately 7.5 kilometers south west of Faridabad Industrial Area and 5.4 kilometers south west of Ballabgarh Industrial Area. It is an interior village with 6 wards. The village has a total area of 4500 bighas. Its total population is 3000 and has a total number of 400 households. The different castes like Jats have 170 households, Brahmins are in 50 households, backward castes have 80 households while Dalits have 100 households. The village has 1 Angadwadi and 1 Government-run middle school, 1 post office, 1 adult education centre, 1 gram sevak samiti, 1 Nehru Yuva Kendra and 1 dairy.

Agriculture and Livelihood The main livelihood sources in the village are agriculture and milk. The village has a total cultivable area of 800 acres. The soil is clay and sandy in equal proportions. In Sagarpur, Kharif crops grown are jowar, bajra, paddy, arhar, maize and dhencha while Rabi crops grown are wheat, mustard and barsam. Some farmers have just started growing carrot, cauliflower and potatoes on a commercial basis.

Issues in Agriculture Some agricultural problems for farmers are those related to short and irregular supply of electricity, pest attacks, high growth of weeds, crop damage by nilgai and irregular climatic conditions like hail storm etc.

Issues Related to Air Pollution All groups correlated health problems such as breathing problems and weakening of eye sight to pollution from nearby factories such as gas plants and brick kilns. Apart from health, pollution from brick kiln had also affected wheat crop. To cope with pollution the village panchayat planted nearly 50,000 trees like Neem, Sesame, Pipal etc.

Name of Village	—	Sahupura
Name of Block	—	Ballabgarh
Name of District	—	Faridabad

Socio-Economic Profile Village Sahupura is located approximately 7 kilometers south west of the Faridabad Industrial Area and 3 kilometers North-West of Ballabgarh industrial area. It is a roadside village with total population of 5000 in 325 households. It is a Jat dominated village along with other castes such as Brahmin, Muslim, Potter, Carpenter, Washerman, Dalit and Barber. Nearly one fourth of the population in village Sahupura is Muslim. The village has a Government middle school and two Anganwadis. There are no factories within the village, though HUDA has acquired 150 acres of land for housing with compensation paid far below market rates. In this village about 100 households are involved in agriculture, which live in the 'Sarpanch' cluster.

Agriculture and Livelihood In this village, around 650 acres of land are under cultivation with 30 tractors and 50 tube wells. The main crop grown is wheat though a number of vegetables is also cultivated. In this village 55 per cent of the people work in Ballavgarh and 45 per cent are involved in agriculture of which, 10 per cent do share cropping. Nearly 60 per cent of income of this village is from land and livestock while 40 per cent comes from labour.

Issues in Agriculture Irregular supply of electricity affected irrigation, which hampered crop production. There was no timely supply of fertilisers and pesticides. Stray cows from Ballabgarh and nilgai attacks caused damage to the crops. White ants destroyed plants from their roots. At present parthenium weed is harmful for human beings and for livestock. Also smoke emitted by thermocol factory from Malerna also affects crop fields of Sahupura.

Issues Related to Air Pollution Air pollution is felt through the wind, which comes from the north west i.e. Malerna and Ballabgarh. Not only mango and guava orchards have been affected by pollution but also crop yield has reduced considerably. The vehicular movement is another culprit. It leaves a thin layer of dust and smoke on the leaves. It prevents crops from growing and leads to loss of crop production.

Doable's The villagers suggested that it is important to improve the supply of electricity. They also said that HUDA should be prevented from taking over cultivable land and agriculture development officer should make regular visits to the village and meet all farmers rather than a preferred few. A people's organization in the village known as Jan Kalyan Samiti, which does community work also decided to take up agricultural issues with the local agricultural department.

Name of Village	—	Sahapur Kalan
Name of Block	—	Ballabgarh
Name of District	—	Faridabad

Socio-Economic Profile Sahupur Kalan is 10 kilometers from Ballabgarh and 17 kilometers from Faridabad and has 168 households. A hamlet named Bhutipura adjacent to the village is also a part of the village Sahapur Kalan though this study only relates to the hamlet Sahapur Kalan. It is a multi-caste village where Brahmin, Rajput, Potter and Dalit live. The dominant caste is that of Brahmins.

Agriculture and Livelihood In village Sahapur Kalan, livelihood of people depends on farming, livestock, labour and salaried jobs where 70 per cent are engaged in land cultivation and livestock raising, 20 per cent in salaried jobs and 10 per cent in selling labour. Income-wise, about 18 per cent of income comes from land and livestock, 62 per cent from job and 2 per cent from labour. There are 400 acres of cultivable land, 30 tractors and 50 tube wells in the village. The Agra canal that commences at Okhla and goes to Agra irrigates about 150 acres of land.

Issues in Agriculture Plant disease was the most important issue in agriculture followed by crop damage by blue bull, irregular electricity supply affecting irrigation, untimely supply of fertiliser and increasing price of agricultural inputs. Other common issues were yellowing of leaves, blackening of crops and curling of leaves. Damaging weeds are on the increase and crop disease is sreading faster than before. Certain variety of paddy and wheat are more vulnerable to pest attack.

Issues Related to Air Pollution There are no local factories in the village but polluted air from factories at village Chandawali blow through the village.

The village groups suggested that there should be six sub-canals from Agra canal for irrigation purposes. The blue bulls need to be kept in sanctuary and prices of fertiliser, pesticide and seed need to be controlled.

Name of Village	—	Sohtai
Name of Block	—	Ballabgarh
Name of District	—	Faridabad

Socio-Economic Profile Sohtai is 6 kilometers from Ballabgarh and has an adult population of 2252 with 450 households. Sohtai is a Rajput dominated village in which Rajputs are settled in 2 Muhallas, Jat in one Muhalla and the other caste in another Muhalla.

Agriculture and Livelihood In Sohtai, 75 per cent villagers are dependent on agriculture and 25 per cent on non-agricultural activity. Income-wise, nearly 80 per cent of the income is derived from agriculture and 20 per cent comes from labour. Nearly 600 acres of cultivable land is there in the village. It has 200 tube well and 22 tractors. Wheat is the main crop with paddy introduced 4 years back. Soil is 'domoth' and water availability is normal.

Issues in Agriculture Cultivators perceived agriculture to be less profitable due to constraints like irregular supply of electricity affecting irrigation, pests and weed attack on crops, crop damage by nilgai and rats, irregular availability of fertiliser and pesticide and pollution from brick furnace.

Issues Related to Air Pollution There are two brick kilns in the village, one established 10 years back while the other was around 3 years back. Both cause local pollution. These kilns operate throughout the year excepting rainy season and lead to black powder deposit on the leaves. Land of the village is also affected by pollution from Chandrawali factories. So also health in the village is affected with increase in respiratory problems.

The village groups suggested that maintenance of regular electric supply was their priority suggestion. Critical support from the agriculture department is required for timely supply of seeds, pesticide, fertiliser and rat poison. The village groups also suggested that polluting factories should be prohibited within the village.

Name of Village	—	Sumper
Name of Block	—	Ballabgarh
Name of District	—	Faridabad

Socio-Economic Profile Village Sumper is at a distance of 6 kilometers from Ballabgarh. Sumper has an adult population of 1714 women and men with households numbering 320. There are different castes residing in the village such as Rajput, Harijan, Brahmin, Barber, Dhimar, Saini, (traditionally vegetable cultivators), Baoria and Gadaria (as livestock rearer). It is a Rajput dominated village with 160 Rajput families residing in two clusters, which are the landowners. There are a total of five clusters, other castes residing in three different clusters.

Agriculture and Livelihood Around 50 per cent of the villagers depend on agriculture and the rest 50 per cent on other activities. Non-agricultural work includes 5 per cent regular job, 10 per cent trading and 35 per cent labourers. Nearly 50 per cent of the income originates from farming, 5 per cent from livestock rearing, 15 per cent from regular jobs, 15 per cent from labour and 15 per cent from petty business. The cultivable area in the village is around 600 acres with 'domath' soil. There are 20 tractors and 30 tube wells in the village. Wheat is the main crop. Vegetables are cultivated on a large scale for water is available and marketing is easy. Share cropping is not that common in this village.

Issues in Agriculture Amongst constraints described by farmers of Sumper was irregular electricity supply, a major one that affects irrigation of fields. The other constraints were crop disease, insect and weed attack, animal attack and pollution. Most of the crop diseases were new and prevalent in improved variety of seeds.

Issues Related to Air Pollution Polluted air blowing from North-South direction carries pollutants from Ballabgarh and Malerna and adversely affects cauliflower, coriander and mustard. A layer of smoke in the fields prevented sunrays from reaching the crops and abated in growth of pests.

The farmers suggested regular supply of fertilisers and pesticide. Elders in the village were of the opinion that the old variety of crops such as barley, maize and millet should be encouraged since they required less water, fertilisers and no pesticide. They would also help in protecting ground water level and improving soil condition whose quality had been reduced.

Name of area   — Uncha Gaon
Name of Municipality   — Faridabad Complex Area
Name of District   — Faridabad

Socio-Economic Profile The village is located approximately 4.3 kilometers south east of Faridabad Industrial Area and 1.3 kilometers south east of Ballabgarh Industrial Area. It is a roadside colony. Unchagaon has a total population of 8000 with a total number of 800 households. Municipak colony of Unchagaon has 1 Government-run senior secondary school.

Agriculture and Livelihood The main livelihood sources in the village are agriculture, livestock rearing and services. Unchagaon has a total cultivable area of 400 acres, in which vegetables are cultivated for commercial production. Unchagaon has clay soil. In Unchagaon, Kharif crops grown are jowar, bajra, maize and vegetables such as lady's finger, bottle gourd, chilli, pumpkin, cucumber, bitter gourd, kakri and melon. In Rabi, vegetables grown are brinjal, cauliflower, cabbage, radish, carrot, spinach, wheat, methi, coriander, sarson ka sag and mustard.

Issues in Agriculture Some major issues in agriculture in Uncha Gaon are water problem, which is increasing with water level falling, crops are becoming more prone to pest and weed attacks, nilgai and stray cows are damaging crops and industrial pollution is damaging crops.

Issues Related to Air Pollution Black fumes from nearby thermocol factories get deposited on standing vegetable crops like cauliflower, tomatoes and spinach. Maximum impact of pollution was felt during the nights of October and November. Crop height and yield are affected for the last 4 to 5 years. There is increase in incidence of TB, asthma, heart attacks, cancer, cough and cold. Fumes waste the flowers of brinjal, tomato and chilie due to ashes falling on them.

4 A Comparative Picture of Pollution, Agriculture and Livelihood in Varanasi and Faridabad

4.1 Introduction

Varanasi and Faridabad are the two districts selected for the study on the impact of air pollution on agriculture and livelihood in urban and peri-urban areas. Results emanating from the study of the villages have been documented in two forms: one, a report distilling peoples perception on a wide range of issues having bearing on agriculture, livelihood and pollution, in particular air pollution; two, case studies of five villages in the two district. This Chapter will deal with a comparative picture of people's perception on the issues at hand.

Varanasi and Faridabad are located in two vastly contrasting settings both politically, economically and agro-ecologically.

Varanasi is in the largest and one of the poorest states of India, namely, Uttar Pradesh, and then in the poorer part of the state, Eastern Uttar Pradesh. Uttar Pradesh is predominantly an agricultural economy and is well endowed in terms of environmental resources. It receives one of the highest rainfalls in the country, particularly Eastern Uttar Pradesh where Varanasi is located. It is fed by two of the major river systems in the country, that of the mighty Ganges and Yamuna. Varanasi is blessed with fertile alluvium soil and is the home of all major religions in the country. Green revolution has caught on in western Uttar Pradesh only; the rest of the State generally hinging on to traditional agriculture. With Green Revolution catching on, poverty is the hallmark of Varanasi.

Haryana is one of the newer and smaller states of India. It was curved out of erstwhile Punjab in 1966. Faridabad is a major industrial center in

Haryana. Haryana and Faridabad lie in the heartland of green revolution. It is one of the richer states in India. Faridabad receives lesser rainfall than Varanasi. Prosperity all round is an unmistakable characteristic of Faridabad.

It is against this backdrop that we will compare people's perspective on pollution, livelihood and agriculture.

We will divide this chapter into nine sections. Section 4.2 will deal with Livelihood. Section 4.3 will be a comparative picture of livelihood and agriculture. Constraints to agriculture and coping strategies will be described in Section 4.4. Section 4.5 will deal specifically on air pollution and its perceived impact on agriculture. In section 4.6 we will look at the institutional issues and section 4.7 to health problems confronting the two communities. Finally we will have a fleeting glimpse of the quality of life in Varanasi and Faridabad.

4.2 Livelihood. Agriculture and Constraints to Agriculture

Patterns of Livelihood

Patterns of livelihood were analysed in villages of both the districts, by men and women separately. In most villages studied in Faridabad and Varanasi, agriculture is very important. From 60 to 90 per cent of the people in different villages are dependent upon agriculture. It is the principal source of food in all the villages. It provides endowment entitlements or exchange entitlements of food to almost all the households. It is the major employer, in both the districts. It is the major source of fodder, either from straw and residues of foodcrops cultivated or from fodder grown as livestock feed, in all the villages studied. It creates effective demand for a whole range of allied ancillary activities like running barbers shops, petty trade, cart pulling, local transportation, tailoring and daily labour.

Rearing of milch animals and selling milk is a very popular economic activity in both the districts, in which men and women all participate, with separate roles curved out in which men sell the milk. Most of the animals in Faridabad villages are water Buffaloes, whereas in Varanasi, there is a mixture of Buffaloes and Cows. The quality of milk produced in the two districts is different and they are used for different purposes, generating different income levels.

Women in most of the villages do not work outside their households, but the rigidity is more pronounced in the villages in Faridabad. At least in one of the Varanasi villages, Tikri, women do go out to work outside their village, but they belong to lower castes and that is rare. Women uniformly have the responsibility of cooking, cleaning, collecting or preparing fodder, collecting fuelwood and drinking water, bear and nurse children. The classical dichotomy presented by "inside home" and "outside home" syndromes operate. Men are mostly engaged in work outside the households and frequently even outside their village, but women are generally given to work inside their households and village. Even when men migrate out of their village, women and children stay back in the villages to take care of their agricultural land, bullocks and ploughs. Generally men are the ones who take household produce for sale, either in the Mandis (markets), nearby towns or within the villages themselves.

Villagers in most of the villages have multiple occupations, particularly men. The range of occupations, outside of agriculture and allied activities, in which people are engaged, is wide, in both Faridabad and Varanasi but farming, livestock rearing, milk selling and working as agricultural labourer, constitute the core occupations. The difference is that the sources of other secondary occupation are different in the two districts. In case of Varanasi, the secondary occupations are more in the service sector and in self-employment ventures in unskilled sectors like running grocery shops, retailing, plying auto-rickshaw and vegetable vending. Whereas in case of Faridabad, secondary occupations are predominantly in the defense services, police force, Government offices, private sector establishments and in skilled or semi skilled sectors. The secondary occupations of the people in the two districts also reflect the huge differences in cultural traits of the inhabitants. Whereas the Haryanvis, particularly the Jhats of Haryana, are known for their ruggedness, valour, gallantry and enterprise, the people of Varanasi are religious, docile, quiet, risk-averse and more attached to their home and hearth.

Changing Role of Agriculture

The composition of the farming community in Varanasi is different from that of the farming community in Faridabad. In Varanasi there is considerable influence of caste in the farming community, in as much as some landowning castes do not cultivate their own land. They lease-out their lands to lower caste people for cultivation, for instance the landowning Brahmin and the Bhumiars. There is, therefore, a high incidence of tenancy in Varanasi. Since law prohibits tenancy in Uttar Pradesh, most of such tenants are concealed. In Faridabad on the other hand, there is no such pronounced trend. There are only class differences designated by size of land holdings. The landowners are also cultivators.

The practice of agriculture in the two districts, mirror a broad similarity but there are basic differences. In case of Varanasi, the impact of Green Revolution is a later phenomenon and is just about caught on. In Faridabad villages, Green Revolution is present in its full-blown form and they are probably entering the "yellowing phase of Green Revolution". In both the places, agriculture has become high external input agriculture, more so in Faridabad. The cost of cultivation has also increased in both the districts but it is much more perceptible in Faridabad. Faridabad villages have also experimented with the "Yellow Revolution" with disastrous consequences but none of it are seen in Varanasi villages (may be again reflecting risk averseness). There is a shift towards cash crops in both the districts but the move is distinctly visible and more down the road in the villages in Faridabad.

In all the villages, the position of agriculture as a means of livelihood, with all its merits and contribution to community life, is on the decline (see box 4.1). In Varanasi and Faridabad, the principal reasons for this phenomenon are increasing constraints, rising costs and greater risks involved in farming; alternative employment opportunities have become available in manufacturing and service sector, and acquisition of agricultural land for non-agricultural purposes, such as construction of houses, industrial and commercial establishments. In Faridabad villages, fragmentation of land due to land-man ratio deteriorating and reluctance on the part of the present generation to soil their hands are additional reasons why acceptability of agriculture as a source of livelihood is on the decline.

There is a clear difference in the approach to agriculture between the older generation of farmers and the younger generation per se in Faridabad. The young villagers in Faridabad, who are educated (the literacy rate in the villages is as high as 90 per cent in some cases), the costs and labour and boredom involved in farming is not worth the cause. They perceive agriculture more from a commercial perspective than anything else. The older generation perceive agriculture as a link between land, livestock, family life and quality of life. To the older generation and women, agriculture provides food, fodder and fuel; land under agriculture is a potential collateral to raise loans; agriculture is a source of employment, for the landless and women; it is a source of income from the sale of cash crops and vegetables. They also see agriculture as a possible hedge against unemployment of their children. This difference in inter-generational perspectives is not there in Varanasi, possibly because the level of industrialisation in and around the Varanasi villages and literacy levels (around 52%) are much lower.

Box 4.1
Reasons for Decline in Agriculture in Faridabad and Varanasi

Varanasi	Faridabad
Increasing Constraints (Power Shortage, No Extension)	Increasing Constraints (Power Shortage, No Extension)
More Pests	New Pests
More Insects	New Insects
More Weeds	New Weeds
More Risks (Erratic and Uneven Rainfall, New Diseases)	More Risks (Erratic and Uneven Rainfall, Sudden Climatic Change)
Alternative Employment available	Alternative Employment available
Conversion of Agricultural land for Construction	Conversion of Agricultural land for industrial and construction purposes
Increasing Costs and falling Returns	New Diseases
	Fragmentation of Land
	Younger Generation unwilling to soil hands
	Falling Returns

Farming Practices

Throughout the villages farming practices have changed. There are three agricultural seasons in Faridabad, namely, Kharif (Monsoon), Rabi (Summer) and Jayad (Winter). In Varanasi there are only two cultivation seasons, namely, Kharif and Rabi.

Box 4.2
Cropping Pattern and Crops Grown in Varanasi and Faridabad

Varanasi		Faridabad		
Kharif	*Rabi*	*Kharif*	*Rabi*	*Jayad*
Paddy	Wheat	Jowar	Wheat	
(Main Crop)	(Main Crop)	(Main Crop)	(Main Crop)	
Maize	Radish	Paddy	Vegetables*	
Jowar	Cauliflower	Vegetables*		
Arhar	Peas			
Cauliflower	Potaoes/Onion/Garlic			
Brinjals	Jowar			
	Methi (Fenugreek)			

*These include Spinach, Cabbage, Cauliflower, Onion, Bottle Gourd, Bitter Gourd, Snake Gourd, Brinjal, Lady's Finger, Fenugreek, Coriander, Radish etc.

As shown in box 4.2, in Varanasi villages, the main crops are paddy in Kharif and wheat in Rabi, with land under wheat considerably less than land under paddy. Kharif is the principal cultivating season. In case of villages in Faridabad, the situation is reverse: Rabi is the principal cultivating season and wheat is the principal crop, grown for consumption. Paddy is cultivated only on a small portion of cultivable land during Kharif, jowar for fodder being the main Kharif Crop. Vegetables are grown in a big way throughout the year mainly as a cash crop. Livestock rearing and milk vending being a very important activity for the Faridabad farmers, they grow jowar as fodder as their main crop.

Improved *irrigation* has played a significant role in both the districts in influencing the cropping pattern. Whereas in Varanasi, with improved, though deficient, irrigation facilities, farmers have substituted growing millets by paddy and wheat, in Faridabad villages, with better irrigation, farmers have abandoned growing grams and other coarse grains, in favour of vegetable cultivation. Similarly, changing over from manure in favour of chemical fertilisers for replenishing soil nutrients has also impacted upon crops grown. Because of chemical fertilisers, land in Varanasi can no longer support cultivation of barley and gram. According to the farmers in Faridabad, due to use of fertilisers (and pesticides), soil conditions have deteriorated. These changes in cropping pattern have their impact on food security, which we will discuss separately.

The intensity of cultivation is high. In both places there is the system of multiple cropping. For instance, during Rabi season (from Aghan to Baisakh) in some of the villages of Varanasi early varieties of vegetables are grown. Once these are harvested for sale in the market, wheat is cultivated for household consumption. Over the years the number of crops grown has declined very considerably.

Seasonality of Activities

Amidst all the fury of living lives under non-optimal conditions, there is a pattern in the lifestyles of farmers in Varanasi and Faridabad. There is drudgery and hard work but variations as well which may be one of the contributory factors leading to their staunch perseverance. The seasonality of activity in the two districts is shown in Table 4.1.

Table 4.1   Activities Related to Agriculture: A Comparative Picture

Month	Agricultural Activity— Varanasi	Agricultural Activity— Faridabad
Jeth-Asar (15 May-15 July)	Sowing and transplanting Paddy. Harvesting and selling vegetables; harvesting and threshing of wheat, and selling surplus. Sowing jimikand, turmeric, pigeon	Preparing land. Sowing jowar, bajra, paddy, moong, etc. Irrigating vegetable fields. Selling vegetables.

	peas, bajra, some vegetables, and moong. Jimikand is harvested and fields cleared.	
Sawan (15 July-15 August)	Vegetable, maize and paddy are cultivated. Paddy transplantation continues. Deweeding of paddy fields. Sowing soya, maize, moong and jowar, and vegetables.	Deweeding fields. Arranging fodder (women). Sowing of dhencha. Cutting of jowar. Preparing seedlings. Irrigating land, if there is no rain. Applying pesticides and fertilisers. Transplanting and sowing paddy. Planting of vegetables. Selling vegetables. Spraying "medicine" for locusts. Deweeding fields.
Bhado (15 August-15 September)	Deweeding paddy fields. Fertiliser application. Planting of paddy saplings. Sowing cauliflower, bajrar and mustard. Harvesting maize.	Irrigating paddy and jowar fields. Cutting leaves for fodder. Planting vegetables. Harvesting jowar and Stocking jowar seeds.
Kuwar-Katak (15 September-15 November)	Ploughing. Sowing vegetables, pea, and gram. Harvesting bajra and paddy. Harvesting and selling vegetables. Harvesting moong (in Kuwar), maize and jowar. In Kartick, ploughing fields for next season.	Harvesting. Stocking seeds. Preparing land. Irrigating and manuring field. Sowing wheat and other minor crops. Cutting for fodder. Sowing and planting vegetables. Harvesting bajra in Kuwar and drying it for 3-4 days.
Aghan-Pus-Mah-Phagun (15 November to 15 March)	Harvesting of paddy (and threshing, parboiling, pounding and dehusking of paddy), mustard and pigeon pea. Sowing vegetables, grams, peas, wheat, sugarcane and melons. Ploughing field several times.	Irrigating wheat fields several times and spraying of fertilizers. In Aghan, sowing crops and vegetables. Irrigating vegetable fields. Applying fertilisers. Harvesting and selling vegetables.

	Applying fertilisers. Irrigating fields with vegetables and wheat. Preparing field for sowing potato. Tilling of potato fields and second round of irrigation. Second round of irrigation, deweeding and application of urea for wheat. Harvesting and washing vegetables.	Deweeding wheat field. Collecting grass and barsam leaves for fodder. Harvesting of mustard. Small farmers extract oil from mustard. Collecting seeds for further processing.
Chet-Baisakh (15 March- 15 May)	Mainly spinach and ninwa together with other vegetables like gourd, radish, etc. are sown. Wheat and tomatoes are harvested; fertiliser application. Threshing of wheat begins. Winter vegetables, grams, mustard and pigeon pea are harvested. Tilling takes place in onion fields; 2-3 rounds of irrigation applied.	Harvesting, threshing, storing, drying and selling wheat. Collecting fodder. Irrigating fields. Selling vegetables. Cutting of barsam and white jai for fodder. In Chet, preparing field for kharif crop; tilling, ploughing, manuring and irrigating fields.

Notes:
(i) This is a general seasonal calendar as described by women groups and men groups from selected villages of Varanasi and Faridabad. Not all households of all the villages are engaged in all activities mentioned above
(ii) There are different crops and vegetables sown in different villages and the above table provides a general picture of the agricultural activities undertaken in the villages of Varanasi and Faridabad, under study
(iii) Information Source: Field reports of PA researchers on Faridabad and the Draft Interim Project Report on "Impact of Air Pollution on Agriculture in Urban and Peri-Urban Areas"
(iv) The data included in the Box do not include purchase of inputs and marketing of outputs (at least not in most of the cases). Ancillary activities like tending the buffaloes/Oxen for the plough and maintenance of agricultural implements

Information from Table 4.1 paints a landscape where the farmers seem to be very busy. They are engaged in some form of agricultural activity or the other in every month: ploughing, preparing land, manuring/fertilising, irrigating, sowing, harvesting, and threshing, selling etc. These are all true for the farmers of Varanasi and Faridabad. These activities do not however, keep the farming households busy on all the days of each month, particularly where we are dealing with small and marginal farmers predominantly. They are in their fields on some days, which allows them to either work in Government, public sector and private sector establishments and come home to cultivate on the days they are needed, or they take up secondary employment when not working in their fields.

And within these months there are differences in the intensity and pace of the activities. For Varanasi, the principal agricultural season is Kharif and the principal crop is paddy (wheat being a relatively minor crop). Hence farmers are extremely busy during the months from Jeth to Ashar (cultivation and post-cultivation activities) and then again in Kuar and Katak (harvesting and post-harvesting efforts). For Faridabad, on the other hand, the principal season is Rabi and the principal crop is wheat, not paddy (indeed paddy is a very minor crop for farmers in Faridabad). The farmers in Faridabad villages enjoy a rather lean period from Ashar to Bhado, but start having grueling time from Katak when wheat cultivation starts, till Chet-Baisakh, when wheat is harvested. Thus both the principal cultivation season and the principal crops are different for Varanasi and Faridabad. From this it follows, that the nature of the agricultural operations are different for the farmers in the villages in the two districts. For farmers in Faridabad, a lot of effort and resources are expended in irrigation for wheat cultivation, because it is an irrigation-intensive crop. But farmers in Varanasi, the principal crop is rainfed and though there is some irrigation, it is limited.

A noticeable difference is that vegetable cultivation occupies a small space in the life of the Varanasi farmers; it is catching on. But for farmers in Faridabad, they are into vegetable cultivation in a big way and both in range and frequency of vegetable cultivation they are further down the road than the Varanasi farmers. The rapid growth of Haryana's economy and the pace of industrialisation in Ballabgarh have created an environment conducive to vegetable cultivation. Vegetable cultivation is also extremely labour intensive and therefore, leaves little time for the farmers in Faridabad for leisure and recreation. Indeed, some Faridabad villagers indicated their preference for cultivating wheat to vegetable, for one reason that the latter required a lot of labour and personal attention of the farmers.

The seasonal activities also indicate that women in Faridabad and Varanasi have different activity sets. For Varanasi, the months of Ashar and Sawan, are for paddy transplantation. In the post harvest period in Aghan-Pus-Mah-Fagun, women put in a lot of effort in harvesting paddy, threshing, winnowing, parboiling, dehusking and pounding paddy. For the women in Faridabad, these activities are on a low key. Their post harvest activities in Chet-Baisakh is threshing and grinding wheat, though with the advent of *aata chakki* (flour mill) the last activity has become less arduous. Deweeding and helping in irrigation, and preparing fodder for the household animals are additional agricultural tasks handled by women.

4.3 Constraints to Agriculture and Coping Strategies

Constraints

Agriculture faces a large number of constraints. The constraints can be variously categorised. For Varanasi and Faridabad, we have categorised these in terms of problem clusters. Thus for Varanasi, the constraints are: (i) weather related, (ii) weed related, (iii) pest and insect related, (iv) animal related, (v) pollution related, (vi) disease related, (vii) land and external input related, (viii) water related, (ix) infrastructure and credit related and (x) institutional arrangement related. In Faridabad the constraints can be categorised as falling under: (i)weather condition related, (ii) weed related, (iii) pest and insect related, (iv) disease, (v) pollution related, (vi) animal related (viii) infrastructure, (ix) external input and output related. The details are given in Boxes 2.7 and 3.5.

Several features of the constraints in both the districts are noticeable. They warrant a few comments. First, the number of weeds in Faridabad are far more than the number of weed related problems of Varanasi villages, though the number of pests and insects are about the same in the villages of the districts. The number of diseases affecting, farming in Faridabad villages in lower than the number of diseases affecting farmers in Varanasi. It is possible that higher doses and higher frequency of insecticides and pesticides applications in Faridabad Villages may have kept at bay diseases but in the process have also destroyed insects and plants which naturally keep weeds under check. In Varanasi, lesser number of pesticide and insecticide application have allowed insects and plants that keep a check on weeds, to survive but in the process the farmers have lost out in contending with a larger number of diseases. This requires further examination. Second, in almost all the villages, the use of external inputs have risen though more in Faridabad villages than in Varanasi villages, where "modern methods" co-exists with semi-traditional methods of cultivation. In all the villages constraints are posed by external inputs, but with a difference. In case of Faridabad villages, the external input type constraints relate more to high costs involved, non-availability, or non-availability in time, of the inputs, while in case of Varanasi villages the external input related constraints relate additionally to poor quality, spuriousness and adulterated external inputs. The difference in literacy levels and difference in availability of extension services (howsoever poor), have something to do with this variation in the nature of the constraints posed by external inputs. Third, in all the villages, pollution of different kinds—air and water pollution as also pollution caused by use of chemical fertilisers/pesticides/insecticides—are perceived as damaging crops. The impact of air pollution is more evident in Faridabad villages than in the Varanasi villages. Probably the higher level of industrialisation in areas around the Faridabad villages than in areas in the vicinity of the Varanasi villages offer some explanation for this difference. Fourth, in villages in both the districts, there are constraints imposed by State policies, which are matters that lie in the domain of public action. Supplying power to agriculture for free may be good but the uncertainty of power supply has done more damage to crops than the good done by the provision of free supply. Similarly, acquisition of land from farmers for different non-agricultural purposes has made inroads into agricultural production and injected elements of uncertainty in farmers' mind, which is detrimental to higher productivity. Finally, when men and women have separately ranked the constraints in the two sets of villages as in box 4.3, the perception of both men and women are about the same in Varanasi. But in case of Faridabad villages, there are considerable divergences in perception on the constraints. Men rank constraints to agriculture on account of infrastructure and external input related problems as their "worst set of problems", women rank constraints posed by weeds and pests as their greatest problems. Generally though there is a broad agreement and views of both women and men groups of villages in Faridabad and Varanasi are found to more or less convergent in identifying infrastructure and external input-related set of problems as one of the priority set of problems. Villagers everywhere find it difficult to cope with bureaucratic inefficiency and lackadaisical attitude, particularly so, of the extension department and state support system. Sadly enough, such problems have intensified over the years, adding to the worries of the farmers. In consequence all over the two districts, production and livelihoods, returns from and cost of farming, both direct and indirect, have all gone from bad to worse for the farmers, who are at the riskiest end of agricultural production. The second prioritised set of problems for women groups relate to crop disease and the adverse impact of pollution. Problem of crop disease is third in priority for men. Farmers have to cope with a variety of crop diseases, some of which are relatively new while others, which are prevalent since long years. The constraints posed by animal related problems are similarly viewed in all the villages of Varanasi and Faridabad: Nilgais, stray cows and buffaloes are as much a problem in Varanasi as they are in Faridabad.

Box 4.3
Prioritised Problem Indices (PPIs) of Agricultural Constraints/Problems

Problem/Constraint Criteria	Faridabad		Varanasi	
	Rank by Women	Rank by Men	Rank by Women	Rank by Men
1. Weather/Climate and natural resource conditions-related	3	6	5	5
2. Weed-related	2	3	7	6
3. Pest and Insect-related	1	2	3	2
4. Animal-related	4	4	4	4
5. Disease and Pollution-related	6	5	2	3
6. Infrastructure/ External Input-related	5	1	1	1
7. Organization-related	-	7	6	7

Then there are differences in how the farmers perceive the constraints in Varanasi from how farmers perceive problems in Faridabad. Of the 7 set of constraints ranked, organisational related problems posing constraints are ranked low down on the priority list of constraints by almost all the groups, and almost all groups consider that infrastructure/external input related problems present the most important set of constraints. All groups rank pests and insect related constraints as a very important set of constraints as well. That is, in villages everywhere have similar perception about the boundary set of constraints. There are differences in perception with regard to the remaining intervening constraints.

It must be mentioned that the constraints faced by the villagers in the villages in Faridabad and Varanasi, have a seasonal character and not all the constraints afflict the villages all round the year. For instance, power shortage, creating problems in irrigating fields, impact on farming throughout the year. Pest and insects appear only during seasons when individual pests and insects thrive. Unseasonal rains come in Cheth and Baisakh and they are therefore to be tackled only during these months.

Coping Strategies

How do our villagers in Varanasi and Faridabad cope with the volley of constraints to which their principal livelihood is subject? How interesting that in both the districts, the coping strategies of the farmers remain almost the same as would perhaps be in all similar rural areas in India. They are rational villagers who have attempted to sort out the problems which, they can sort out at their level, through personal effort. These include substituting or avoiding crops like chickpeas, sugarcane and lintels, certain kinds of vegetables, and cash crops like mustard, which are vulnerable to pests, weeds and disease; fencing and guarding fields to ward off Nilgais and stray animals; poisoning rats; a tendency or inclination to move out of agriculture; selling off agricultural lands and migrating to other areas; hunting for supplementary sources of income to finance high cost agriculture and/or hedge against crop failures; application of pesticides and "medicines" based on experimentation of fellow farmers and advice from dealers and even hiring generators for ensuring power supply at times when irrigation is needed.

Everywhere farmers perceive and recognise the importance of the state as manifested in the Block Development Officer, the Agriculture Department, Electricity Department, Panchayat Pradhan , Seed Farms and even the Minister in one case. They are aware of the support that their agriculture can receive to their overall wellbeing, but they lament that these institutions are either non-existent or non-functional. They are in some cases far removed from the psyche of the farmers for them to even think of approaching them. Some farmers have tried to bring to the notice of these support institutions/ departments, their problems but to little avail. For instance villagers of Malerna complained about noise pollution to higher-ups in local administration and the factory owners, but that has not helped. There is no forum to listen to their complaint, like the Consumers Redressal Forum or a forum for listening to problems of bank customers and even insurance policy holders. The insensitivity of the support institutions and government functionaries to the problems of the farmers comes out quite convincingly. As a result, in many cases, farmers rely more as the next best alternative, on the traditional support structures like the village money lender, the shopkeeper who sells insecticides and pesticides, the seed supplier and the like, despite the knowledge that they often get spurious and adulterated inputs. The support that farmers get from these sources may be at exorbitant costs (nominal and real) but they are reliable in terms of their timing and flow of supply.

4.4 Impact of Air Pollution

Different kinds of pollutants are present in Varanasi and Faridabad as described by the local village communities. Such pollutants are basically air and water.

Air-pollution is linked to industrial estate in near-by towns of Ballavgarh and Faridabad for villages in Faridabad, vehicular-related, local factory-related. Ballavgarh and Faridabad industrial belts are close to the villages and the smoke and different pollutants from the factories pollute the air and settle on crops and affect crop yield. The air pollution has increased due to considerable increase in the number of vehicles in that area over the last 20 years. This has led to smoke and dust in the air; it has also affected crop production. Pollution through local factories and industrial units like those of brick kilns, thermocol, plastic and cement pipe factories have affected crops and health of both human beings and animals.

Though not related to air pollution yet given the importance of the pollutant, contaminated water from industrial units, found in some villages to have negatively affected crops, health of human beings and animals. Agra canal had polluted water about a year and half back near the village Sahapur Kalan, adversely affecting crops and vegetable cultivation, and both their taste and nutrition level of crops, as described by senior farmers.

The women and men groups in Varanasi villages described the sources of local pollution as industrial estate, industrial units, domestic coal stoves, garbage and vehicular pollution. The women and men groups are aware of selected impacts of air pollution upon village communities. In general, the villagers relate the direct impact of air pollution on their health and less so on crops and agricultural production, unless it is an impact of pollution quite close to their field, like the ones caused by brick kiln near-by. Scientifically, though there is evidence of one to one relations between certain pollutants and certain kinds of disease, weeds, pest attacks adversely impacting on crop yield and growth, the village communities have different perspective in this regard. The PA researchers have observed that the incidence of pest, insect and weed is noticed more in those villages, which are situated in the ozone pollutant zone, where dependence of people on farming is greater than on other means of livelihoods.

On the whole, the sources of and impact of air pollution are perceived more in the villages of Faridabad than in the villages of Varanasi. There could be several reasons for this, including higher industrialisation in areas around and in the vicinity of Faridabad villages, more green cover and natural vegetation, and the overall level of awareness, consequent upon higher literacy rates of the villagers. It is also to be underscored that though there has been general agreement on the pollutants and their impact, in some cases perceptions did vary and there apparently was no reconciliation. For instance in Sahupura, some villagers reported that they have stopped mustard cultivation owing to the damaging effect of air pollution. But others maintained that the damage to mustard was caused by chepa (a pest) and not by air pollution. There are other instances as well.

4.5 Institutional Issues

We have already detailed the institutional issues while discussing the constraints, Suffice it to say for the present that a lot needs to be done.

In Varanasi, dysfunctional or non existent institutional facility for supply of credit; lack of support mechanisms for imparting knowledge of scientific cultivation; lack of knowledge of proper selection and use of seeds; absence of agriculture-related training and extension services, are serious problems. Communication gap between villagers and officials; closing down of the local institutions such as the cooperative centre in one village, due to corruption and bankruptcy are some of the institutional problems farmers face. The absence of any institutional mechanism to ensure that farmers get a fair price, is a big institutional problem. In Faridabad, acquisition of agricultural land by Haryana Urban Development Authority and absence of water in government canal are complex issues. For instance, lack of water in Government Canal has forced farmers to use water from sewage canal. This means that lack of canal water is tantamount to directly channeling pollutants into otherwise fertile cultivable land of farmers, for entirely avoidable reasons. Fertilisers are not released in time by the government and hence crops do not get nutrients at the right time. Supply of Fertilisers is not a problem. Demand for fertilisers exists. There is need to have a bit of co-ordination, driven by institutions and systems. Similarly non-availability of good quality seeds is no mean problem that can be easily rectified by simple changes in institutions.

At present the existing support system is clearly not that responsive. There are no NGOs working in the villages at present and community-based organizations like local Panchayat are not that effective. For instance, in Sahupura village, a people's organization called Jan Kalyan Sewa Samiti was registered in 1992 by village youth, which looks after village cleanliness but not into agriculture issues. Similarly, such an organization exists in Sahapur Kalan as well, though not involved in agriculture-related work. The villagers of Kadhaoli actually complained that lower level of awareness amongst the villagers about pollution and low level of literacy, is partly due to the absence of any NGO. Land-related conflicts are very common and people take help of the local magistrate's office or the police. The sub-divisional agriculture office at Ballavgarh caters to 192 villages.

Visits of extension officials are not regular and do not help in solving local problems. Visits of officials from other relevant departments are few, which benefit generally the well to do farmers who have some clout, while others remain unaware of such visits. Indifferent attitude of the local state departments related to agriculture is a major stumbling block. Irregular supply of fertilisers and pesticides results in crop loss. Credit agencies also do not respond adequately to farmers' needs.

It is sad that after expending so much of political effort and resources to amend the Constitution to provide for Panchayats as the third tier of Government, these institutions have not taken roots, either in Varanasi or in Faridabad. And that too even after five years of arduous march on the road. While, the Panchayat Samitis in the villages in Varanasi were perceived not fulfilling their duties/responsibilities fully. The village Panchayats apparently could not perform as they received only 60 per cent of the total grant from the government. Most of the villages in the jurisdiction of Panchayat Samities were deprived of the various schemes undertaken by it. People in general felt that the Samities had failed miserably to perform its welfare activities in the areas of providing medicines in wells, conducting immunisation programme, family planning and other related programmes. If it is argued that Panchayat Samities are enjoined to perform any of these are functions, it only shows that the people are not even made aware of the roles of Panchayats.

The picture is not very different in Faridabad. For many villages, the local Panchayat is not that effective for lack of funds, and hence authority to take decisions. For instance, there are cases where these institutions were unable to take action, even where formal complaints were lodged with them, against pollutants being released by brick kilns in one case and themocol factory in another. While, probably because of higher literacy levels, the people in Faridabad do not have high expectations of the Panchayats, they do perceive that these important local level institutions operate only at the margin in their scheme of things.

Thus in both Varanasi and Faridabad villages, people look upon Panchayats as important local level institutions. In Varanasi these institutions are seen as not fulfilling their developmental roles, in Faridabad people have negative experience with Panchayats in fulfilling their protective or regulatory roles.

4.6 Comparative Picture of Health Status

The health status of the people in the villagers in the two districts is a matter of concern. In both the district villagers suffer from a large number of diseases, symptomatic of a situation where health infrastructure is as bad as the institutional arrangement for supporting farming. The common diseases in the villages of Varanasi are cough and cold, T.B., skin diseases, eye diseases, stomach problems and "enlargement", malaria, jaundice, gastritis, frequent body ache and mild fever among the elderly, asthma, frequent brain fever amongst children, increased incidence of pimples all over body of children, diarrhea, ear disease and pus flowing out of ear amongst children and sores on different parts of the body.

In Faridabad in addition to the diseases mentioned by farmers in Varanasi, skin disease, TB, cancer, heart attack, nose burning, headache and dizziness, blood pressure, heart ache, "baye" (arthritis), pain in appendix, allergy, weakening of eye sight, dengue (incidence for the first time two years back), acidity and pain in stomach, young women have serious cases of leukorria (white discharge), pain in the waist, tooth decay among children and breathing problems.

It is apparent that there are three kinds of health problems in the villages. *One*, diseases are chronic in nature which, affects large sections of the population irrespective of season; *two*, diseases are specific and affects only some people, and *three*, diseases which, afflict the general population during certain times of the year, and relate to public health. For example, stomach problems, cough and cold, asthma, body ache and fever, headache, arthritis, weakening of eyesight and sores are general ailments and are chronic in nature. We may categorise them as falling in the first category. heart attacks, T.B. and problems of blood pressure and such other ailments are people specific, and fall in the second category. Then dengue, diarrhea, malaria, jaundice and brain fever come in spurts and during certain parts of the year to which the entire population is vulnerable, is in the third category. These diseases are matters that lie in the domain of public health. The fact that all three kinds of diseases affect people in our villages points to the need to have preventive health interventions, promotive health interventions and curative health care. While curative health can be left to individual initiative, the need for preventive and promotive health forming the core of health strategy of the government in both the States (and Districts) can not be overemphasized.

If the two lists are considered, then it seems that the number of diseases perceived to be present in Faridabad are higher than the number of diseases in Varanasi. Since there is no data on how many people actually suffer from these diseases, we can not conclude either way whether people in one district are healthier than the other.

Villagers' perceptions about the status of health in the two districts are different and they are significant. Farmers in Varanasi articulated the problems of child health better than problems of the general population. While the villagers in both the districts gave a long list of diseases, the Faridabad villagers related the deteriorating health status to industrialisation, which the farmers of Varanasi did not. The villagers of Varanasi emphasized the problems of mal-nutrition amongst children, which was seemingly not an issue in Faridabad, probably because of better food security situation in the latter than in the former. The farmers in Faridabad villages related some of their ailments clearly to increased local emissions, villagers in Varanasi related some of the diseases suffered by them to consumption of foodgrains grown by using chemical fertilisers and pesticides and changes in climatic conditions. This may be partly explained, prima facie, by the fact that the level of industrialisation is higher in and around he villages in Faridabad than in Varanasi. The differences in villagers' perceptions in the two districts are tabulated in Table 4.2.

4.7 Quality of Life

Quality of life is an issue which did not figure on a separate plane of inquiry but every element discussed is interwoven into the fabric of quality of life: pollution, crop yields, health, employment, livelihood and food security. In both Varanasi and Faridabad hamlets, majority of people felt that the process of local industrialisation has helped them by offering more employment, a larger market for their product and labour, and higher prices for their products such as milk and food grains.

Industrialisation and urbanisation has been perceived as a mixed blessing by the villagers in both Faridabad and Varanasi, despite differences in their respective levels of industrialisation. In the villagers' perception, in the villages in both the districts, though they have derived benefits, the costs of industrialisation have been quite considerable. The women groups were particular, in the villages of Faridabad, in pointing out the costs of urbanisation and industrialisation.

But the costs were different. For villagers in Faridabad, the major costs were scarcity of grazing land, loss of village identity, health complications due to pollution, increased cost of living, congestion, decline in the quality of social environment and damage to crops. They have also indicated deterioration in the quality of water and actions by Haryana Urban Development Authority to acquire land, particularly agricultural land, for non-agricultural purposes, as costs of industrialisation. The major costs perceived by villagers in Varanasi were cultivation of market-driven commercial crops, higher doses of external inputs and high accompanying risks in agriculture and higher pest problems. Migration of men in search of jobs created special problems for women. Increased pollution negatively affecting food and fruit production and falling sanitation standards with industrial waste and higher crime rate, were the other problems for Varanasi.

Table 4.2 Comparative Status of Health in Faridabad and Varanasi Villages

Health Status in Varanasi	Health Status in Faridabad
1. Both women and men groups have listed a number of diseases and illness, which they and their children suffer from, but did not relate them to any phenomenon.	1. The community members are aware of the deteriorating health conditions in recent years with rapid industrialisation at local level.
2. There is increased incidence of cough, respiratory disease, stomach-related disease, skin disease, eye infection, TB and malaria, most of them amongst children.	2. There has been a rise in respiratory diseases, breathlessness, burning sensation of throat, eyes and nose and headache, with higher incidence in winter months and higher pollution.

3. Many children are also victims of nutrition related problems.

3. The health expenses have gone up for local communities since home medicines hardly work. Health of their livestock is also affected by air pollution causing indigestion, diarrhea and death.

4. The village groups relate the diseases to a mix of factors such as increased use of fertilisers and chemicals in agriculture causing stomach-related disease/illness and respiratory diseases due to change of weather.

4. It was not difficult for many village groups to draw one to one correspondence between local emissions and physical discomfort and illness, which they suffer.

5. However, not many villagers perceive any impact of air pollution causing serious impairment of health.

5. Community perspective on general health status indicates adverse impact of air pollution apart from other common diseases.

4.8 The "Do-able's" at Varanasi and Faridabad

Local community members of the villages under study suggested the following "do-ables" for overcoming the problems and constraints faced by them and also in improving their quality of life. The "do-ables" are those, which the local village communities expect the government to do. The emphasis is more on the expected role of the government and its agencies on three generic issues as follows.

In Faridabad

- Offsetting Damages from Industrialisation and Urbanisation.
- Enhancing Agriculture Support System.
- Strengthening Agricultural Policy Support.

In Varanasi

- Can ensure supply of electricity for irrigating crops and make loans available at concessional rates.
- Can provide fertilisers and medicines at subsidized rate.
- Can take disciplinary actions against manufactures of duplicate medicines.

If we look at the generic "doables" that the villagers in the two districts have basically outlined, there are clear differences, and they reflect the state of poverty and the state of the art in agriculture in these districts. In case of Faridabad, villagers want the Government to deal with variables that are exogenous to cultivation, whereas villagers in Varanasi want the Government to deal with variables that are endogenous to farming, mostly to do with direct *external* inputs. In Faridabad, the villagers perceive the role of the State in its regulatory dimension more important, whereas the villagers in Varanasi look upon the State as their benefactor —State as the benevolent State. Because farmers in Faridabad are relatively better off, in several respects, not only in monetary terms, they need less of State patronage to manage their inputs, whereas the poorer farmers, more disempowered, in several respects, need the long arms of the State to sail through. In both cases, the dimension of an appropriate policy framework has been spelt out.

4.9 Concluding Remarks

Thus in the villages of Faridabad and Varanasi, there is clear indication that villagers are aware of the ongoing industrialisation, benefits from it and the costs involved. The increase in the number of pollutants, and levels of pollution have been recognised. Their effect has been identified, but it has been linked better with health status of the villagers than with the health of their agriculture. The constraints to agriculture have been identified and they too are perceived as increasing. In general there are similarities in the perception of the villagers about the broad trends both in Faridabad and Varanasi. There are differences, which are space specific and gender specific. These differences have lessons for policy making.

5 Lessons from the Field

Some of the lessons on participatory field research learnt from the field are described below. The lessons have been grouped into the following generic clusters.

- Approaching Community for Participatory Research.
- Listening and Learning from Local Community.
- Pace of Research.
- Conducting Field Inquiry.
- Research Methods.
- Quality of Research.
- Handling Sensitive Issues.
- Managing Community Expectations.
- Making Research Culturally Compatible.
- Approaching Research Frontiers.
- Communication with Scientists and Other Stakeholders.
- Research Team's Commitment and Spirit.

Approaching Community for Participatory Research

The process of rapport building with local communities can itself be a lengthy process. In the project under discussion, it was often done through helping them in their daily work or by chit chat sessions. To offset the extractive nature of such participatory research, considerable emphasis was placed on proper rapport building with communities and explanation of objectives of field research. The issue of air pollution was not introduced directly as one of the objectives of the project so as not to bias the community members in any way. The issue was rather broached in an indirect manner in which the community themselves sought to include such issue either in their description of agricultural constraints or in their discussion on health issues or through assessment of impact of industrialisation and urbanisation. Once the community flagged such issues they were taken up for further discussion.

Different ways were adopted for approaching study areas. One way was to approach and discuss the research objectives with the village head for informing the villagers before visiting a village. This route was adopted especially for villages visited in Faridabad. Whereas, in Varanasi, the goodwill of BHU in and around its near-by villages proved useful. Perceived credibility of research by the community was also high and doubts and misunderstandings were easy to handle. Rapport building took more time in the Faridabad villages, especially in the first round. The successive rounds were much easier for the PA researchers since the process of establishing mutual trust building had already started. The follow up rounds of field visit helped in re-establishing trust and in convincing the villagers. Such follow-up rounds were taken up on the advice of the villagers.

Lending a Helping Hand

The local community members of Varanasi and Faridabad are busy throughout the year in different activities. One way is to help them in their work as much as possible so that some kind of goodwill is created between external PA researchers and the community members/group interacted with. It helps in better understanding and rapport building by contributing to the daily lives of local communities. Many PA researchers working for the project gained acceptability and credibility by lending a helping hand, which helped in establishing channel of communication. The community, in the process was able to appreciate the research objectives. To start with, in the first round, the PA researchers were naturally treated with suspicion often mistaken for officials from government (to conduct survey on some sensitive issue such as land holding), officials from income tax or from other departments such as HUDA to acquire land. Gradually, such initial apprehensions transformed into constructive relationship based on mutual trust and faith. This improved with lending a continuous helping hand and also re-visiting the village communities. The bonds got stronger for some PA researcher who also followed up by writing correspondence and sending photographs to community members. This helped in maintaining contact with community and also in gap filling.

In the sowing/harvesting season, the villagers are too busy. Some of them also take leave from city work to come and sow/harvest. To learn about the issues at the sowing/harvesting stage it may become important to interact with the community. It was an opportunity for PA researchers to help the farming community through sowing and thus improving rapport. Lending a helping hand for sowing helped in breaking ice and in sustaining constructive relationship.

Listening and Learning from Local Community

From the perspective of participatory field research, It is important that any local community is able to appreciate the objectives of participatory research and agrees to participate. Gaining the confidence of local community members is important without raising their expectations. Hence, listening to the community, learning and going by their advice helps in acceptability of field research by the community. For the fieldwork in the current project, it was the community, which emphasised staggered visits to learn about the relevant issues important to the project. Based on their advice the PA researchers undertook round of visits to the community in those seasons/periods, which the community indicated. Most of their problems related to the different seasons. Such seasonal dimensions of the problems were best understood when the community was visited in different seasons as and when their problems occurred. One lesson learnt in the process was that projects of such nature need to have enough flexibility to incorporate community advice into its working plan for achieving best results. This also helps in strengthening relationship with communities.

Another lesson learnt is that participatory research concerning intricate issues in biophysical resources is better learnt and appreciated as and when they actually occur. They are understood properly through joint observation and mutual discussion with participants who are repository of indigenous knowledge. For example issues about the impact of air pollution have a large "visual" element (for example leaf injury or dust deposit) and learning about it as it happens enables the community to explain it vividly on the spot. Sometimes "words" or verbal explanation fall far short of the "visual" features of the problem under description. Joint walks and joint observation in team of community members and PA researchers also helps.

Pace of Research

Need to Keep Pace with Community

Local communities practising agriculture in urban and peri-urban areas are busy throughout the year either in farming or in off-farm activities. They have little time for outsiders. However, time is an important factor for both "breadth" and "depth" of research. And process-oriented participatory research needs time, for it involves community analysis, probing, observing, recommending and cross checking.

The project earlier specified limited time for field research but later responded to a staggered approach for field research since community interactions were required to be periodic, based on a day-to-day approach. On a daily basis, such interactions were for shorter duration since community members, both women and men, were busy in different activities. As a result, once rapport was built and the research objectives were explained to the community the research progressed at a slow pace given the willingness of community members to participate. All field research was undertaken during the time, which community found convenient. One important lesson from project experience is that it is important to keep pace with the way of life of the community. The community members under study are generally busy and have little time in each season when they have a particular set of crops and activities to focus upon.

Pace of PA research varied from hamlet to hamlet. So also were the perspectives, which differed. Such perspectives differed group-wise from same hamlet. Appreciating and learning about different perspectives helped in furthering research investigation and bringing improvements in the checklist.

Conducting Field Inquiry

One lesson learnt from the project is related to the way of doing in-depth topical research in agriculture. For best results, it is important to conduct staggered rounds of investigation in line with the agricultural crop cycle. It enables local community to describe their priority issues to external facilitators in a concrete manner (when the problem is actually arising in the field). It also helps in mutual discussion amongst themselves in order to identify related issues and arrive at group consensus. As for external PA facilitators, such staggered rounds of research help in locating problems/ constraints faced by local communities, better appreciation of live issues by closely observing them, understanding them and conducting in-depth discussions with the concerned community/groups.

Both exploratory and topical research investigations were important from the PA research point of view. At the exploratory stage, the PA researchers learnt of different issues raised by village groups of women and men following the initial checklist of issues proposed under research investigation. Based on community participation, this checklist grew in size. Whereas, in-depth research on selected issues was conducted during the phase of topical research, which generally followed the exploratory phase. Such topical research issues were based more on the mental checklist of local community.

Since results from field research get utilized as building blocks for policy analysis and advocacy in the area concerned, it is important to both "broaden" and "deepen" such research base. Policy suggestions and advocacy emerging from a broad research base are robust and can withstand validation by different local groups, which face similar set of problems, though not identical.

Research Methods

Appropriate methods are important to be sorted out and flexibility in methods is the crux of participatory approaches. In PRA there is a large inventory of methods, some of which are more effective than the rest for certain topics. Appropriate selection of methods and their sequencing help in better research investigation and analysis. One such method found effective for field research was the "live sample" method used by local communities to introduce samples of their problematic pest/weed/disease. This method helped in having a "visual" and a concrete base for discussion. It created an environment for group participation, made more participants interested in analysis of their own problems and helped in generating discussion on a large number of related issues, which the participants deliberated in an exhaustive manner.

One lesson to be drawn from such field experience is related to the limits posed by verbal discussion. Some of the problems and the impacts of the communities under reference are quite complicated and verbal description may not be that effective. In such cases the best approach to adopt is to examine the real conditions/objects and initiate discussions on that basis. The two methods, which proved effective for such field inquiry are the "visual" live sample method and the joint field walk and field observation method followed by group discussion. The joint walk and observation was undertaken by the PA researchers along with the community members to observe selected agricultural problems/constraints in the field and have discussions on them.

Quality of Research

Quality of interaction is often more important than quantity since it is a major determinant of the kind of information generated in the process. Quality of information is related to the process adopted. The better the rapport building and interactions with the community, better is the quality of information and its depth. The local groups of women and men felt free to speak and express their views and engage in deeper analysis. They were withdrawn in their approach when community-based interactions were superficial with greater degree of "extractiveness", which happened with one research team at Faridabad. Such superficial interactions with community not having much confidence in external researchers, affected research output.

Whereas, daily interactions with the community members and research rounds succeeding the first round helped in assuring the community about the perseverance, transparency and seriousness of PA researchers and helped in building mutual relationship, confidence and trust. One lesson is that sheer quantity of information adds small value to the process of field research and findings.

Handling Sensitive Issues

Some topics are sensitive in nature and questions related to them can create conflict. In such cases it is better that the local groups themselves find opportunities to raise issues without leading questions being posed by PA reserachers. Topics like industrial pollution are sensitive, often having a background of local grievance and conflict. In such cases it is important to get a feel for issues, which are locally sensitive and observe local "do's" and "don't's" on discussion of such issues. This helps in making the environment congenial and constructive. Potential areas of conflict need to be handled with care and tact. Hence, the way questions get posed is important. The locally sensitive issues, if tackled in non-threatening manner, help community/groups to think in a positive way and look for scope for resolving conflict and do mutual confidence building.

Managing Community Expectations

Responding to issues raised by community is not always easy. The community members are generally looking for suggestions/solutions to numerous issues that they raise. The PA researchers at Varanasi were in continuous dialogue with the scientists (engaged in doing research under the project) and gave them feedback from the local community. For villages in Faridabad, some issues were communicated to the government departments concerned especially the state extension department and the research station, while some social issues were also resolved through mutual discussion of PA researchers and local groups. Since much remains to be done in terms of enabling community members to do their networking, it can be taken up as an objective in the following phase of the project, if any.

Making Research Culturally Compatible

Often women in conservative societies have problems talking to men PA researchers where the latter, may not find it convenient to do so given the cultural framework and vice versa. Hence it is important to devise ways to offset gender bias in such projects, for example by having research teams with both women and men. Offsetting 'dominance' of men and village leaders is another aspect, which require tact and deft handling by PA researchers. One lesson in this regard is from Varanasi where men influenced priority issues of women group/s. This implies further probing in Varanasi for arriving at analysis of difference. For gender perspective it is important the women and men groups are consulted as different groups.

Approaching Research Frontiers

New areas/topics of field research, which are generally under-researched require time for carrying them out. Technical issues in research require a fair amount of technical knowledge and understanding. It also suggests building communication channels with technical person/experts for better clarification of technical problems raised by community/group/researchers. Scientific briefing of PA research team (assuming that many have social science background) is important for those technical topics, which are relatively new and normally not researched. Only those scientists would be effective in briefing social science researchers who can de-mystify the terms, jargons, processes, technical facts etc. and explain them in simple language. In this respect, those scientists are better equipped who are in a position to appreciate community perspective in field research. It is not always easy to find many of them as and when required for supporting PA research.

Communication with Scientists and Other Stakeholders

Conversations with scientists and experts with ongoing PA researchers help the latter with social science background to clarify technical concepts and are mutually beneficial. It helps the latter to appreciate scientific and environmental issues raised by local community members and to see the relationship from an inter-disciplinary point of view. It also helps them to raise further queries on the subject with the local communities. Such participatory interactions can help in exploring the societal basis of scientific and environmental phenomena. The village groups are also benefited in the process. This actually happened at Varanasi where the scientific team took considerable interest in participatory research output and responded to some of the results from field inquiry.

Research Team's Commitment and Spirit

It is important to have team members with commitment and quality and not necessarily large teams for doing research. Such large team becomes a crowd in the village and can disturb participatory process and findings. A small committed and dedicated team is in a better position to establish effective communication with focus groups as compared to a large team, which helps in quicker coverage of groups/communities. It is spending quality-time with the community where appropriate attitude and behaviour help in establishing effective communication with local groups. This, if done by a small team, helps in improving the quality of relationship over time. One lesson is to have a small number of field researchers and to provide a reasonable time for field research.

6 Methodology of Participatory Approaches

Introduction

In Chapter 6 the methodology of field inquiry has been described and discussed. At the beginning of the chapter the research objectives have been listed and the participatory approaches methodology has been outlined. The other discussions in the chapter relate to ways adopted for approaching field for conducting both exploratory and topical rounds of inquiry. This is followed by description of approaches adopted by researchers during field research for participatory interactions/discussions with different village groups. Criteria for village selection is also described and different participatory methods used by the researchers are listed. The field research cycle highlighted in the chapter helps to provide contextual meaning to the kind of results arrived at. A generic checklist of issues used by the researchers is described for appreciation of the aspects emphasised during research. The checklist of issues is at best indicative for an open-ended participatory inquiry into the subject of research. The research team and their code of conduct are also listed. This is followed by a brief description of dissemination of lessons from the field and a list of project outputs.

6.1 Research Objectives

The objectives of participatory approaches to field research were as follows.

- To learn about livelihood strategies of local communities in the study areas and the role and importance of agriculture and agricultural practices.
- To learn about the constraints in agricultural production and the ways in which such constraints are addressed by local communities.

- To explore whether air pollution is recognised by local communities as a serious constraint on agricultural production.
- To learn about the health status of local communities.
- To learn about the step/s suggested, if any, for overcoming constraints in agricultural production (including both adapting agricultural practices and/or lobbying for changes).

6.2 Participatory Approaches Methodology

The methods and principles of PRA were adopted for field research. Though the principles of PRA were foremost in deciding the basis of research, the approach adopted for field research can be best described as that of Participatory Rapid Rural Appraisal (PRRA). "PRRA is a variation of PRA, which is widely used where information is required by external agencies but must be expressed by the communities themselves in their way and with their own emphasis." It is at variance with RRA, which is "one off", extractive exercise and, in which, the outsiders gather information in the shortest possible time based on their own interpretation of the outcome. Whereas, PRRA is used to acquire just enough information of local women and men groups' own experience and own analysis of the situation for selected development intervention, development planning or as inputs into policy research. Like RRA, PRRA is a "one off" and extractive exercise but is based on the process of participation in which people themselves define their priorities. PRRA often provides illustrative community views, which can then be used as an entry point for more intensive participation in the long run. PRRA, however, is different from PRA, in which local communities are empowered to take control of their own development (UNDP:1996).

6.3 Approaching Field

The participatory research had two parts—exploratory and topical. For all field sessions it was important to start with warming up and rapport building exercises with local communities and groups. This also included explanation of project objectives without raising expectations. The exploratory part of the field research entailed a general approach towards exploring the objectives of the project under consideration, e.g. household livelihood strategies in general or livestock problems, in general, for having overview of the issues concerned. This was followed by topical round of research for focusing on selected issues e.g. constraints related to wheat crop or issues in women's empowerment and livelihood. There was the additional issue that people did/did not recognise, such as the direct and indirect impact of air pollution on crop production. The fact that it was mostly "invisible" meant it was not widely recognised as a "problem" by farmers. The participatory approaches (PA) researchers inquired into ways and means of how people dealt with agricultural constraints (which captured some of the pollution-related issues) and also how they viewed their health and other problems and what factors they thought were causing them. The PA researchers generally avoided asking direct questions on impact of air pollution but facilitated it through indirect means like asking about their agricultural disease, health related issues and historical trends in yield/crop disease/human health etc. Joint walk to the fields with villagers, discussion of agricultural constraints, livelihood and health and other issues, most often, led to discussion on selected impacts of air pollution. However, the village group/s did not always relate them to air pollution as such, including the obvious ones like chlorosis/leaf injury in crops.

6.4 Participatory Ways for Field Inquiry

One general way adopted was to inform the Pradhan/village head about the village visit and then meet the local community members to introduce oneself about the visit and project objectives. This also helped in selection of hamlets and village groups and in deciding time for meeting-discussions. One other way was to get acquainted with community members like school teachers, anganwadi workers and other members and greet them and chat with them, clarify objectives of village visit, get introduced to other members, request for time for discussion etc. Another way adopted was to visit a village, meet any random group of people by the roadside/village tea stall, introduce oneself and initiate general discussion about the visit and the village. Base line data from block office or village officials helped in appreciating the background of the village under consideration.

6.5 Criteria for Village Selection

The PA researchers in consultation with the scientific research team selected villages/hamlets in Faridabad and Varanasi for conducting participatory research. Such selection was refined and improved upon in consultation with village communities. In this context, some general criteria applied by PA research team for selecting villages in Varanasi and Faridabad are described below in box 6.1. For specific criteria related to each village see Annexes 'A' and 'B', in the chapters on Varanasi and Faridabad. The "muhallas/bastees" or wards from each village were selected depending upon the availability of agricultural community, socio-economic characteristics of wards etc. The selection of wards within each village was mainly based on discussions with the village groups.

Box 6.1
Criteria for Selection of Villages

- Selecting villages with different pollutant levels as per scientific data e.g. selection of villages in areas with high ozone pollutants but relatively low levels of other pollutants and vice versa.
- Selecting villages close to sites where scientists are measuring pollution levels and impact on crops.
- Coverage of villages where different communities are residing for example Hindus, Muslims, Dalits etc.
- Coverage of villages located near roadside and those further interior.
- Selecting villages of both large size and small size.
- Coverage of villages located close to industrial sites and other areas.
- Coverage of villages having farming as main livelihood and those not having so.
- Coverage of villages based on proximity to industry-types.
- Selecting villages with and without local industry.
- Selecting villages based on different crops.

The researcher carried out participatory field study as per the project objectives. Gender was mainstreamed into the fieldwork for approaching local community and learning about different perspectives (e.g. those of men/woman, old/young, rich/poor) in the communities undertaken for the study.

6.6 Participatory Interaction/Discussion

Often a common meeting place was chosen as the place for meeting-discussion with village groups. Much emphasis was given on the process of rapport building with the village groups. When farmers/agricultural labourers were busy in the field, methods such as 'do-it-yourself' and 'lend-a-helping-hand' were applied by PA researchers followed by group discussions and interactive sessions held near the field. Sometimes the farmers suggested alternate dates and time, which were then respected. PRA-type sessions were mostly held in separate groups of women and men, though some mixed groups were also interacted with.

Some discussions were also held in the homes of the village participants. Men groups often made drawings and other "visuals" and wrote down their analysis on chart papers with colour pens, some drew on the ground, which were recorded by the facilitators in their note book. One women group also did "visuals" and wrote down their analysis and shared it with their fellow participants. This helped in cross checking. The participants also did oral analysis, which was recorded by the facilitators.

The participatory methods as shown in table 6.1 included participatory mapping, group discussion, joint observation, seasonal analysis, problem tree, benefit-cost matrix, criteria listing, scoring/ranking, do-it-yourself, lend-a-helping-hand, time line/oral history, card sorting, poverty grouping, wealth grouping, visual assessment, live samples etc. PRA chart 6.1 provide field illustrations of participatory mapping and daily routine of village Adityanagar, Varanasi. PRA chart 6.2 show symbols used by village group of Harijan Basti of village Tarapur, Varanasi.

Box 6.2
Participatory Methods in Village Tikri, Varanasi

"The senior participants segregated themselves into a separate sub-group and recapitulated their experience as per my request. The very first theme that we discussed was the major events that occurred during past. The participants started to narrate events from 1940 onwards....The discussion started from the earliest decade—1940's. I had to write down the gist of discussion because none of them volunteered to write or to sign. However, all of them were going through the sheet."

Source: Field Report on Varanasi by Meera Jayaswal

The PRA principles of learning, listening, respecting community views formed the crux of the research. PA researchers adhered to learning from the field and sharing results.

A relatively new tool was applied in this study, which helped it immensely. In this "live sample method" the villagers took help of live samples of pests/weeds/crop disease to explain, prioritise and rank their agriculture-related issues. The method drew a lot of interest from participants and made for intense discussion. It also helped in adding related samples from more participants and contributed towards in-depth probing.

Table 6.1   Participatory Process and Methods

Topics	Participatory Methods Used for Field Study
Livelihood practices and food security	Group discussion, pie-chart, flexi-interview, time line, seasonal calendar, livelihood mapping, sketch mapping, livelihood assessment.
Role of Agriculture	Oral history/time line, group discussions, seasonal calendar.
Gender Issues	Focus group discussions with women groups and men groups, case study, interview, seasonal calendar, daily routine.

Wealth grouping	Card sorting, wealth mapping, group discussion.
Farming systems crops yields, inputs, output, returns etc.	Group discussion, time line, seasonal calendar, joint walk, social cost-benefit matrix, time trend, live sample method.
Agricultural constraints, prioritising constraints, coping strategies, suggestions and others	Group discussion, criteria listing, scoring/ranking, oral history, problem tree analysis, flexi-interviews, joint walk, observation, venn diagram, seasonal calendar, live sample method.
Role of industrial area and its impact	Group discussion, criteria listing, oral history, cost-benefit assessment.
Impact of air pollution on agriculture and human health, coping strategies and suggestions	Group discussion, joint walk, time line, time trend, seasonal calendar, listing, impact matrix, joint observation.

Note: Methods also included warming up, rapport building, key informant analysis, do-it-yourself, lend-a helping hand, chit-chat, secondary data analysis, reporting back to villagers, participants' cross-checking, peer-group cross-checking, sharing results in seminar-cum-workshop

For each of the two study areas, Faridabad and Varanasi, the PA researchers prepared field reports based on their research results. The main points of the reports were drawn for by PA research coordinator for preparing the synthesis report on community perspective on agriculture and air pollution as described in the project objectives, as above. The PA research coordinator also pursued field research from time to time to validate select findings and also to add value to the sum total of findings.

The PA researchers discussed findings and shared lessons in peer group meetings from time to time, though informally. On 2nd April, a seminar-cum-workshop was held by the project at India International Centre, Delhi where most researchers presented and discussed their field reports in considerable detail in presence of external resource persons and other collaborators. Selected results of the study were also disseminated to some local officials, at Faridabad and Varanasi, involved in local development and extension activities, research etc.

Table 6.2 provides an overview of the villages at Varanasi and their constituent hamlets covered during field research, the names of PA researchers, criteria for selection of village, number of community members interacted with and the methodology/approach adopted by PA researchers for field inquiry.

Table 6.3 provides an overview of the villages at Faridabad and their constituent hamlets covered during field research, the names of PA researchers, criteria for selection of village, number of community members interacted with and the methodology/approach adopted by PA researchers for field inquiry.

Table 6.2   Villages of Varanasi—Participatory Inquiry in 14 Villages

Name/s of Village/Hamlet	Name/s of Researcher/s	Main Criteria for Selection of Village	Number of Local community Members Interacted with	Methodology/ Approach Adopted
(1)	(2)	(3)	(4)	(5)
Village-Seer Govardhanpur Hamlets Visited —Not Available	ActionAid, India	In this village, the pollutants are Nitrogen Dioxide and Sulphur Dioxide.	Total no. of participants—32 No. of Women 15 No. of Men 17	Prior to fieldwork, visits were made to initiate rapport and ask community for their convenient time. Transect and farm sketches were undertaken in all the villages, as an effective source for rapport building. Starting point was what people had to share with us. The team encouraged the farmers to do "visuals" for their responses. Transect, trend analysis, Venn diagram, seasonality exercises and focussed discussions were common methods applied in the field study.
Village-Sarai Dongre Hamlets Visited —Not Available	ActionAid, India	In this village, the pollutant is Ozone.	Total no. of participants—24 No. of Women 12 No. of Men 12	

(1)	(2)	(3)	(4)	(5)
Village-Lohta Hamlets Visited —Not Available	ActionAid, India	In this village, the pollutant is Ozone.	Total no. of participants—46 No. of Women 24 No. of Men 22	The initial day of each village visit was spent in visiting the village chief's (gram pradhan) house for informing her/him about the visit and getting an 'overview'. In the absence of the chief, some other village key informants were approached at random for learning more about their village.
Village-Chitaipur Hamlets Visited —Not Available	ActionAid, India	The village has Nitrogen Dioxide and Sulphur Dioxide pollutants.	Total no. of participants—21 No. of Women 10 No. of Men 11	
Village-Nathupur Hamlets Visited —Not Available	ActionAid, India	The village is near the Diesel Locomotive Works where Nitrogen Dioxide and Sulphur Dioxide are at moderate levels.	Total no. of participants—33 No. of Women 24 No. of Men 9	
Village-Tikri Hamlets Visited —Mahato Basti, Dherabir Basti, Yadav Basti and Chowdhurian Basti	Sudipta and Meera	In this village, Ozone level is highest with high levels of Nitrogen Dioxide and Sulphur Dioxide.	Total no. of participants—93 No. of Women 55 No. of Men 38	

(1)	(2)	(3)	(4)	(5)
Village-Maraon Hamlets Visited —Maurya Basti and Patel Basti	Sudipta	The village is near the Diesel Locomotive Works where Nitrogen Dioxide and Sulphur Dioxide are at moderate levels together with high levels of Ozone.	Total no. of participants—62 No. of Women 18 No. of Men 44	On the basis of such primary information, the hamlets got selected, which were later visited for field inquiry. Both women and men groups were explained the objectives of the field visit and were requested a convenient time for meeting. Efforts were made to inform as many villagers as possible. Separate interactive sessions were held with women and men groups at different locations. The methods for participatory interaction included participatory group discussion, flow chart, seasonality, chappati diagram, historical transect, participatory mapping, pie chart etc.
Village-Maheshpur	Sudipta	The village has high levels of Nitrogen Dioxide and Sulphur Dioxide pollutants.	Total no. of participants—11 No. of Women 6 No. of Men 5	
Village-Adityanagar Areas Visited—Municipal and non-municipal areas	Meera and Nabin	Direct exposure to vehicular pollution is high—100 per cent of the village area on the Pachkosi Road.	Total no. of participants—31 No. of Women 12 No. of Men 19	The initial day of village visit was spent on an exploratory survey of the village for an overview. Meeting with village chief (gram pradhan) on the very first day and informing him of the visit and its objectives was important. Such information helped in sampling of

(1)	(2)	(3)	(4)	(5)
Village— Karamanbir Hamlets Visited —Khanpokhari, Bichlapura and Karamanbir Brinda	Meera and Nabin	Direct exposure to vehicular pollution is low—25 per cent of the village area is on the roadside.	Total no. of participants—38 No. of Women 20 No. of Men 18	hamlets and groups and also in arriving at a mutually convenient time for holding discussions. The following days of village visit were spent in investigating the range of research issues by using 'visual' and 'verbal' methods of PRA.
Village—Nuaon Hamlets Visited —Bhumiar, Patel, Rajbhar and Harijan	Meera and Nabin	Direct exposure to vehicular pollution, relatively higher than village Adityanagar— 90 per cent of village area on Mughalsarai–Mohansarai bypass road.	Total no. of participants—79 No. of Women 43 No. of Men 36	
Village—Tarapur Hamlets Visited —Bhumihar, Brahmin, Harijan and Rajbhar	Meera and Nabin	Direct exposure to vehicular pollution is very low— this village being an interior village; there is existence	Total no. of participants—44 No. of Women 22 No. of Men 22	

(1)	(2)	(3)	(4)	(5)
		of Ozone pollutant.		
Village-Chandpur Hamet Visited— Patel Basti	Hema, Sunita and Meera	The village has relatively higher levels of Nitrogen Dioxide, Sulphur Dioxide and Ozone.	Total no. of participants—15 No. of Women 9 No. of Men 6	The initial day of the visit was spent in identifying key informant/s, clarifying objectives of field visit and having general discussion on village and its historical background. One major hamlet of the village was selected for field inquiry. Participants within a hamlet were selected at random for interacting both with groups and individually, on different issues related to field inquiry. A combination of PRA methods such as participatory mapping, time line, wealth ranking, seasonality and problem listing and scoring were used to generate information and also to cross check results.
Village— Navampur Kalan Hamlet Visited— Yadav Basti	Hema and Sunita	In this village, the pollutants are Nitrogen Dioxide and Sulphur Dioxide which are at moderate level.	Total no. of participants—6 No. of Women 2 No. of Men 4	

Table 6.3 Urban and Peri-Urban Areas of Faridabad—Participatory Inquiry in 14 Areas

Name/s of Village/Hamlet/Municipal Town	Name/s of Researcher/s	Main Criteria for Selection of Area	Number of Local community Members Interacted with	Methodology/Approach Adopted
(1)	(2)	(3)	(4)	(5)
Village—Kadhaoli Hamlets/Wards/Communities Visited— Aat Bisa, Barah Bisa	Sudipta and Meera	The village is located about 6.5 kilometers West of Faridabad industrial belt and 4.3 kilometers West of Ballabgarh industrial area.	Total no. of participants—21 No. of Women 14 No. of Men 7	For approaching local communities in any village, the village chief (Sarpanch) was generally visited and apprised of the purpose of the visit. Generally a common location like a village tea stall was selected and people from different hamlets were invited for group discussion. The local people were briefed about village visit of PA researchers and the objectives, and, they, in turn, provided 'overviews' about the village, based on which hamlets got selected. In the selected hamlets sessions were held separately with both women and men from different socio-economic groups. The village women and men described
Village— Sagarpur Hamlets/Wards/Communities Visited— Patyat Muhalla and Hawelli	Meera and Sudipta	The village is located about 7.5 kilometers South West of Faridabad industrial area and 5.4 kilometers South West of	Total no. of participants—25 No. of Women 8 No. of Men 17	

(1)	(2)	(3)	(4)	(5)
Walle Muhalla **Municipal Town**—Uncha Gaon Hamlets/Wards/Communities Visited—Saini Muhalla		Ballabgarh industrial area. The village is located about 4.3 kilometers South East of Faridabad industrial area and 1.34 kilometers South East of Ballabgarh industrial area.	Total no. of participants—17 No. of Women 4 No. of Men 13	their village/hamlet and the changes that had taken place in agriculture, farming system, cropping pattern and growth in local industries. Current issues in the village were also discussed covering their livelihood strategies, agricultural constraints, health, infrastructure, support system and costs/benefits of industrialisation and urbanisation. Participatory methods, both 'visual' and 'verbal' like historical transect, time line, 'seasonality' analysis, criteria listing and scoring, group discussions and semi-structured interviews were used for interacting with the villagers.
Village—Baroli Hamlets/Wards/Communities Visited—Jhandwa Muhalla	Sudipta, Meera and Madhumita	The village is located about 1.6 kilometers South East of Faridabad industrial area and 2.6 kilometers east of Ballabgarh industrial area.	Total no. of participants—22 No. of Women 8 No. of Men 14	

(1)	(2)	(3)	(4)	(5)
Village— Pali Kasba Hamlets/Wards/Communities Visited— Sahpuriya Patti	Sudipta and Meera	The village is located about 7 kilometers North West of Faridabad industrial area and 8 kilometers North West of Ballabgarh industrial area.	Total no. of participants—22 No. of Women 10 No. of Men 12	On the initial day of village visit, the local village Chief (Sarpanch) was approached and briefed about the project, its nature and objectives of field visit. After a first round of introduction with a small group of villagers, larger groups of women and men were met for group discussion and PRA exercises. In all
Village— Chandawali Hamlets/Communities Visited— Brahmin and Jhat	Sudipta, Meera and Bratindi	The village is located about 5 kilometers from Faridabad industrial area.	Total no. of participants—109 No. of Women 64 No. of Men 45	
Village—Malerna Hamlets/ Communities Visited— Yadav, Brahmin and Jath	Bratindi	This village is located 4 kilometers from IOC*; is exposed to Faridabad industrial pollutants and also	Total no. of participants—90 No. of Women 50 No. of Men 40	

(1)	(2)	(3)	(4)	(5)
		has local factories causing pollution.		villages, group discussions were held in common meeting place (baithak). Objectives of field visit were clarified on several occasions and an attempt was made to cover maximum people involved in agricultural activities. Farmers with different size of land holdings and from castes and age groups were approached for discussion. Participatory methods used for field visit were time line, do-it-yourself, group discussion, pie chart, case study, criteria ranking, field transect and Venn diagram. Information generated was cross-checked and additional information was added as and when required. Two rounds of inquiry were undertaken for each village. The third round of field visit was spent in planning village level workshop of local farmers with the local extension department of the government.
Village—Sumper Hamlets/Communities Visited—Brahmin, Rajput and Harijan	Bratindi	This village is located 4 kilometer from IOC*; is exposed to Faridabad industrial pollutants and also is exposed to vehicular pollution.	Total no. of participants—72 No. of Women 50 No. of Men 22	
Village—Sohtai Hamlets/Communities Visited—Brahmin, Rajput and Harijan	Bratindi	This village is located 5 kilometer from IOC*; is exposed to Faridabad industrial pollutants and also has local factories causing pollution.	Total no. of participants—37 No. of Women 21 No. of Men 16	

(1)	(2)	(3)	(4)	(5)
Village— Sahapur Kalan Hamlets/ Communities Visited— Rajput, Jath and Gujjar	Bratindi	This village is located 5 kilometer from IOC* and is exposed to Faridabad industrial pollutants.	Total no. of participants—97 No. of Women 42 No. of Men 55	
Village— Sahupura Hamlets/ Communities Visited— Yadav, Jath, Brahmin and Khati	Sudipta and Bratindi	This village is located 4 kilometer from IOC*; is 7 kilometer South West of Faridabad industrial area; 3 kilometer North West of Ballabgarh industrial area; is exposed to vehicular pollution; and also has local factories causing pollution.	Total no. of participants—92 No. of Women 32 No. of Men 60	

(1)	(2)	(3)	(4)	(5)
Village— Jharsainthly Hamlets Visited— Not available	Indian Social Institute, New Delhi	This village is about 11 kilometers from the Faridabad town, on the main Delhi-Mathura road; it lies on both eastern and western side of the Grand Trunk road.	Total no. of participants— Not available	Entry into the village was made either by contacting the village Pradhan (Sarpanch) or by direct contact. Group and individual discussions and interviews were held in the village for gathering perceptions from different socio-economic segments of the village. Discussions were held with big, small and marginal farmers and with non-farming groups such as agricultural labourers, lower income groups, Dalits, landless backward class etc. which included both women and men groups.
Village— Jhajru Hamlets Visited— Not available	Indian Social Institute, New Delhi	This village is located about 13 kilometers from Faridabad town and about 3 kilometers in the interior north east of the Grand Trunk road.	Total no. of participants— Not available	

(1)	(2)	(3)	(4)	(5)
Village— Piyala Hamlets Visited— Not available	Indian Social Institute, New Delhi	This village is located 15 kilometers from Faridabad town on Delhi-Mathura highway and 3 kilometers interior (North—East) from the highway.	Total no. of participants— Not available	

* IOC stands for Indian Oil Corporation

Source: Based on Field Reports of PA Researchers, Faridabad

6.7 Field Research Cycle

Though it wasn't envisaged at the beginning, village-based field research had to be staggered for creating space for topical round/s of field inquiry in the same villages. Rounds of field research had to be adjusted towards the crop cycle. The local groups/communities expressed their opinion on the issue of agricultural constraints and strongly suggested that researchers re-visit them for topical investigation related to agricultural constraints such as weed/insect/disease/others. According to the local group/s, agricultural issues under investigation were better discussed with such issues occurring naturally over the crop cycle. It becomes easy to identify and learn about any agricultural constraint as and when it actually occurs. The local group/s emphasised the crop cycle for better discussion of issues and their "visual" description.

Staggered visits over the crop cycle helped in generating in-depth discussion and in understanding of issues based on visual analysis. It also helped in rounds of follow-ups where some samples of pest and disease were provided to the PA researcher who in turn passed them on to the scientists, especially at Varanasi for further analysis. See Box 6.3.

Box 6.3
Showing Sample of Brown Plant Hopper from Paddy Field to Scientists

"The team proceeded to Chitaipur to collect sample of brown plant hopper in paddy. Since it occurred for the first time, most farmers were not able to differentiate between blight and this. Further, the crop has been harvested by many farmers. After visiting six sites of the standing paddy crop, we were able to collect the sample..... Later, the team met two senior members of Department of Mycology and Plant Pathology in Banaras Hindu University, to obtain technical clarification on the constraints reported by the farmers. All the samples collected during the field work were shown to them for their opinion."

Source: Field Report, Varanasi by ActionAid, India

Staggering the pace of research also meant the same researchers re-visiting villages for topical research. Hence the project was required to start small and adjust the village visits of research teams and follow a process of periodic interaction with local communities over their range of crops that they cultivated.

There was a team of 2 core researchers visiting both Faridabad and Varanasi with inputs from additional researchers including research teams from ActionAid India, Centre for Social Development and Indian Social Institute. For those researchers with full time jobs it became difficult to spare time for the project and adjust to the actual research schedule. Hence the two core researchers Meera and Sudipta, were able to provide a kind of continuity to the project by adjusting their research time with the crop cycle. They conducted both exploratory and topical research in Faridabad and Varanasi. Bratindi was also engaged in both the study areas though for Varanasi, she did the research on behalf of ActionAid, India as a mixed team as compared to Faridabad where she did it on her own as an individual researcher.

Though many other individuals expressed their interest to be associated with field work they could not be contracted for various reasons such as their individual office not allowing leave for research, personal problems, prior commitments etc. The field research required enough space for flexi-time of both local community members and PA research team. Since most researchers were employed in full time jobs or busy with other projects, the project was required to make continuous adjustments to align researchers' time with the crop cycle of the study areas. Some of the field reports were written in Hindi while others, though written in English, were presented and discussed in Hindi during peer group meetings.

6.8 Checklist of Issues

The PA researchers were provided with a preliminary checklist and some briefing on field methods by the PA Research coordinator. A preliminary checklist (given in box 6.4) used by PA researchers, for field inquiry, had 4 broad parts.

- Household livelihood strategies and role of urban/peri-urban agriculture.

- Constraints and problems of agriculture production.
- Ways in which such constraints and problems were being addressed.
- Impact of industrial area on household livelihood and agriculture.

Box 6.4
Preliminary Checklist

- Household livelihood strategies and role of urban/peri-urban agriculture:
  - households, socio-economic groups, livelihood pursued;
  - livelihood in agriculture, description, nature, options, seasonality etc.
  - description of seasonality and its impact of living and quality of life;
  - role of agriculture to local economy;
  - role of other livelihood to local economy;
  - farming system practices, crops grown, land related issues, quality of land related issues;
  - land ownership, labour involved, wages, agriculture inputs;
  - loans, indebtedness—patterns/sources—influencing factors;
  - crop rotation, yield trends, transport, markets;
  - return from agriculture/other livelihood;
  - income, savings, expenditure, food patterns;
  - institutional structure.
- Constraints and problems of agriculture production:
  - list of constraints/problems;
  - inter-relationship amongst them, if any;
  - prioritising constraints/problems;
  - description of nature of constraints/problems;
  - factors influencing constraints/problems;
  - impact/damage of constraints/problems;
  - estimates.
- Ways in which such constraints and problems are being addressed— coping strategies, support system, institutional support and networks, local initiatives and impact, measures undertaken/ to be undertaken, role of local government/other organisations like CBOs/farmers' organisa-tion/NGOs/others—any farming system research/experiment—its nature and its role—ways of coping with the damage.

- Nature of action/s being taken to overcome constraints/problems—what future action is feasible, where, when and how, by community, groups, by panchayats, by government, by NGOs and relative success of different strategies mentioned.
- Impact of industrial area on household livelihood and agriculture—benefits and costs—impact on agriculture including impacts on health or other aspects of life.

6.9 Selection of Research Team

Selection of researchers for the project was not an easy task. For field research, it required selection of those researchers who were known to have commitments to participatory approaches (PA), analytical skills to explore a relatively new research topic, some writing skills and enough time for doing field research. For the fieldwork, individual researchers were selected with different academic background and experience. During selection of PA researchers, three main aspects were emphasized.

- one was the attitudinal bent of individual researcher implying respect and appropriate attitude towards local communities;
- second was that of having some knowledge and basic skills in participatory approaches; and
- third was the ability to converse in local language with local communities at Faridabad and Varanasi.

Since the research topic was relatively new, effort was made to have combination of both senior researchers with more field experience and junior researchers with relatively less field experience but those eager to interact with local communities and arrive at fresh findings. Three senior researchers and four junior researchers were contracted on an individual basis while research teams from three non-governmental organizations, Centre for Social Development, ActionAid India and Indian Social Institute also conducted field research. One underlying criterion for contracting researchers was to have a mix of different disciplines and background to the extent possible. One specialist prepared the case study volume for the project based on field reports from Varanasi and Faridabad. The following gives some idea about the background of the researchers.

- Hema—studied Anthropology, is carrying out Ph.D. research and is engaged in social mobilisation in watershed areas in Doon valley, Uttar Pradesh.
- Sudipta—studied Economics and marketing, was employed in corporate sector and is trained in man management tools.
- Meera—Ph.D. in Psychology, teaches Psychology to post-graduate students in Ranchi University and is researching in education and health in tribal areas of Bihar.
- Sunita—studied Economics and is engaged in social mobilisation in watershed areas of Doon valley, Uttar Pradesh.
- Bratindi—Ph.D. in Psychology, is engaged in project funding of development activities in Bihar.
- Rai and Prasad—field workers from Centre for Social Development.
- ActionAid team—Damodaran, Alok and Bratindi are engaged in community appraisals and project funding in community development and poverty alleviation in Bihar.
- Madhumita—Ph.D. in Social Geography, is associated with training in gender issues.
- Team from Indian Social Institute—Alka, Tyagi, Janaki and Archana, all have done their Ph.D.s in natural sciences like micro biology, agronomy, plant genetics, nutrition and are working in the social sector for research and training.
- Navin Datta-Banik—Masters in Geology, Diploma in business management, studied air and water pollution and now working in an environmental firm.
- Amitava—Ph.D. in Economics, ex-Executive Director, ActionAid, India, Now Honorary Professor at Institute of Human Development, New Delhi, is researching in participatory development, food security and human rights.

6.10 Code of Conduct for Field

As far as the PA methodology and its principles are concerned, the PA research team, by and large, adopted the following rules for field research. Such field rules are basic to the principles of PRA.

- respecting local communities and building rapport;
- asking community for time and respecting community time;
- clarifying objectives to the community and not raising their expectations;
- having a flexible approach;
- avoiding road-side bias;
- using checklist of issues to be probed and taking notes;
- sequencing participatory methods and cross-checking results;
- reporting back field results to the community;
- reflecting on lessons and embracing errors;
- having team spirit.

6.11 Dissemination of Lessons from the Field

PA researchers provided feedback to the community on research findings and cross checked field output through application of different participatory methods, interacting with different groups at different places and time and also through peer group discussions. One of the PA researchers also exchanged letters with local community members for filling information gaps to which such community members gladly responded. During the period of field investigation, the PA researchers were in continuous touch with the PA research coordinator.

6.12 Research Output

There was a series of project outputs from the field research on community perspectives. It included the following.

- Village reports written by PA researchers on their field visits.
- An interim synthesis report at the meso level on field results from Varanasi and Faridabad.
- A discussion summary on the main field results from Varanasi and Faridabad.
- A final synthesis report at the meso level on field results from Varanasi and Faridabad.
- A report on case studies from selected villages of Varanasi and Faridabad.

References

Mukherjee, Neela (1993), *Participatory Rural Appraisal, Methodology and Applications*, Concept Publishing Co., Delhi.

Mukherjee, Neela (1995), *Participatory Rural Appraisal and Questionnaire Survey: Comparative Field Experience and Methodological Innovations*, Concept Publishing Co., Delhi.

UNDP (1996), *Report on Human Development in Bangladesh, A Pro-Poor Agenda*, volume 3, United Nations Development Programme, Bangladesh.

Appendix 1
Community Outputs—PRA Maps and Charts

Appendix 1 contains selected PRA outputs from the villages of Varanasi and Faridabad as listed below.

Chart 1—Participatory map, village Karamanveer, Varanasi.

Chart 2—Venn diagram, village Tarapur, Varanasi.

Chart 3—Poverty grouping, village Nathupur, Varanasi.

Chart 4—Venn diagram, village Sahupura, Faridabad.

Charts 5A, 5B and 5C—Past, present and future mapping of village Chandawali, Faridabad.

Note on Community Outputs

The original maps and charts made by the village communities of Varanasi and Faridabad have been reproduced. Most of them have legends in Hindi language for which the following note helps in understanding them.

Chart 1—Participatory map, village Karamanveer, Varanasi is a social-cum-resource map of village Karamanveer, drawn by village participants named Kailash Prasad, Annu Prasad and Uma and facilitated by Nabin Datta Banik. The map shows the agricultural fields, houses, ponds, wells trees and village roads.

In Chart 2, a Venn diagram of village Tarapur, Varanasi, the perceived social relationship of villagers with different institutions is shown. The Venn diagram shows the organizations identified by the farmers as part of their agricultural support system such as local agriculture department, irrigation department, power department, local market, Kisan Sewa Kendra (farmers' service centre), bank, panchayat/gram sabha (village council) and cooperatives. The size of circles of local agriculture department, irrigation department and power department, though large are at a distance from the

189

farmer implying the weak support that is provided by them. Whereas, local market, Kisan Sewa Kendra (farmers' service centre), bank, panchayat/ gram sabha (village council) and cooperative are relatively closer though circles are smaller in size. This diagram was facilitated by Meera Jayaswal and Nabin Datta Banik.

Chart 3 shows poverty grouping at village Nathupur, Varanasi done by village women of Nathupur based on different criteria of poverty. The chart shows households at two points of time, one is dated 40 years back and the other is the latest position. Though the conditions of dwellings have progressed from huts to structures made of bricks, the size of agricultural land of individual households has considerably declined over time. The criteria of poor households, which emerged from discussion are those who do not have agricultural land, lone widow and those who do not get employment. The chart was drawn by women participants of village Nathupur by the names of Urmila devi, Sri devi, Jugura devi, Mujla devi, babna devi, Janno devi, Manno devi, Amravati devi, Manbhabati devi and Sushma devi. The village women group was facilitated by Bratindi Jena.

Chart 4 is a Venn diagram of village Sahupura, Faridabad drawn by a group of well to do farmers from village Sahupura. In the Venn diagram, the social relationship of the village with other agencies is indicated. For the group, market is the most important place for them as they procure all their requirements like seeds, fertiliser and pesticides from the market. Another important place is the Ballabgarh grain market, place where they sell their products. Office of the Sub-Divisional Magistrate is another place where they go for settling their land disputes. But it is relatively less important and is at a greater distance. Village Panchayat is of very little importance as it is not capable of rendering much help. Rural bank is perceived as important and the circle is at a relatively close distance. The participants were Jagram, Bachuu ram, Om Kumar Sharma, Meher ChandSharma, Suraj Pal Chowdhury, Sukh Pal Chowdhury, Shiv Singh, Ganj Pal, Maha Singh, Sukh Ram, Swarup Singh, Daya Chand Sharma, Chandrapal and Girija Singh. The Venn diagram was facilitated by Bratindi Jena.

Charts 5A, 5B and 5C depicts past, present and future position of residents of village Chandawali, Faridabad. As explained in box 3.11, Women farmers in Chandawali village have drawn three pictures to depict their weak position against the rapid pace of industrialisation. Chart 5A is depicting the position of village households 40 years back when they owned land, owned bullocks to plough, had good crops, trees, water, well, school for children and used their skills in artisanship. Chart 5B shows their position today, when they have little or no land, agriculture requires tractors, pesticides and fertilisers and their agricultural land is being gradually taken up for industrial units like cement factories, brick kilns etc. Chart 5C shows the position after 40 years, when the women group imagine that they would lose all their land, their fields and their livestock and there would be multi-storey flats in the locality where they and their children will be forced to work as domestic help. The participants of women group were Rachna, Asha, Suman, Uma, Chameli, Kanima and Vilesh. The village women group was facilitated by Bratindi Jena.

Chart 1 - Participatory Map
Village Karamanveer, Varanasi

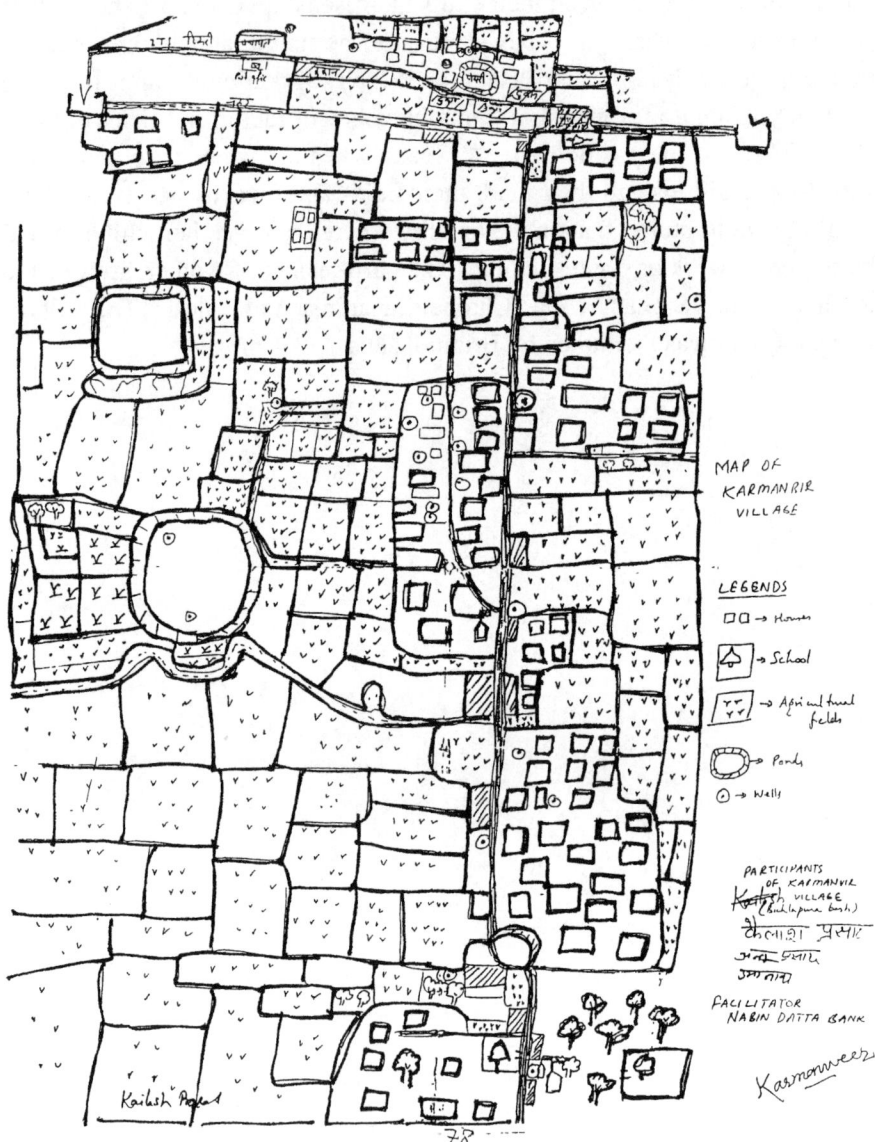

Chart 2 - Venn Diagram
Village Tarapur, Varanasi

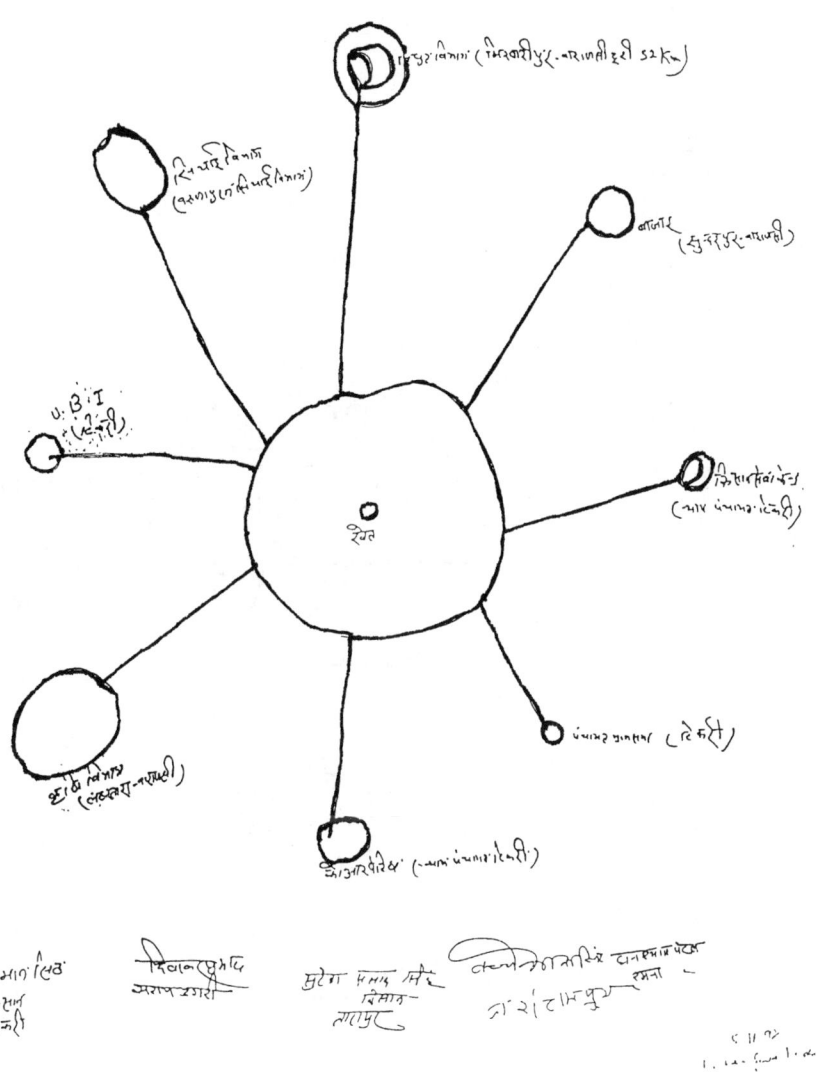

Chart 3 - Poverty Grouping
Village Nathupur, Varanasi

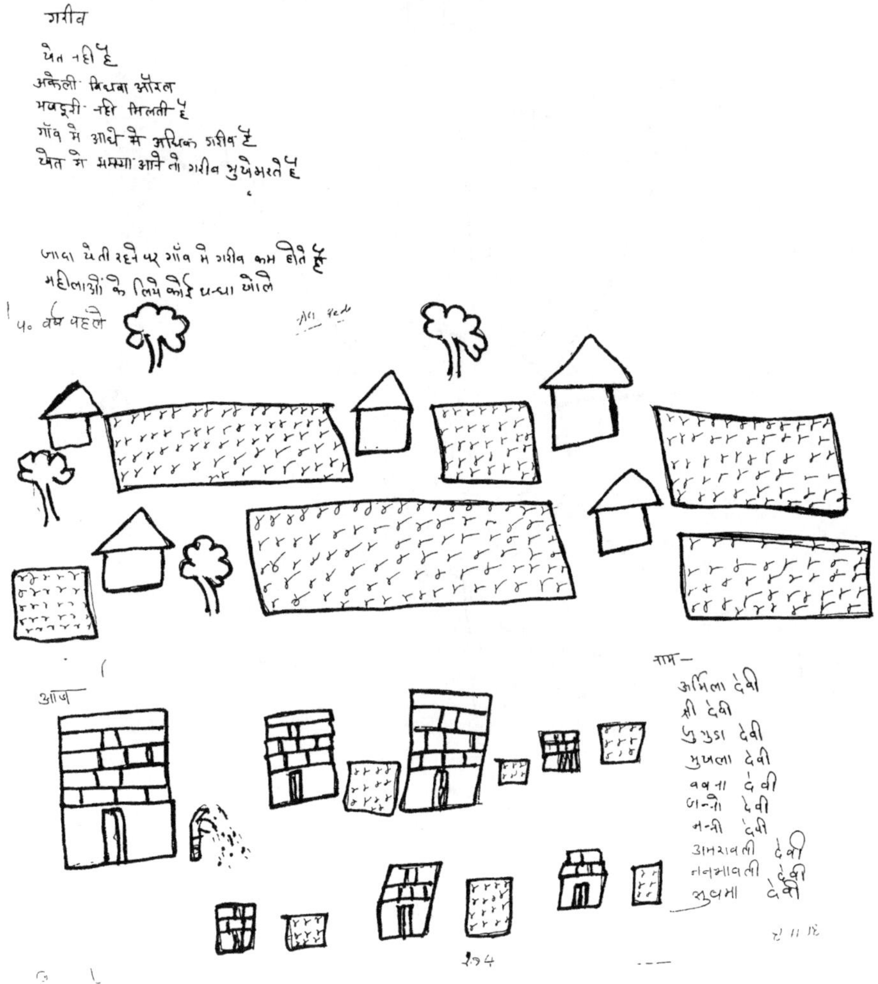

गरीब

खेत नहीं है

अकेली विधवा औरत

मजदूरी नहीं मिलती है

गाँव में आधे में अधिक गरीब है

खेत में मम्मा आने तो गरीब मुखे मरते हैं

जादा खेती रहने पर गाँव में गरीब कम होते हैं

महिलाओं के लिये कोई धन्धा नहीं

४० वर्ष पहले

आज

नाम —

ऊर्मिला देवी
सी देवी
जुगुठा देवी
मुफला देवी
वबना देवी
ब्रानी देवी
नन्नी देवी
आमरावती देवी
ननमावती देवी
जुलमा देवी

Chart 4 - Venn Diagram
Village Sahupura, Faridabad

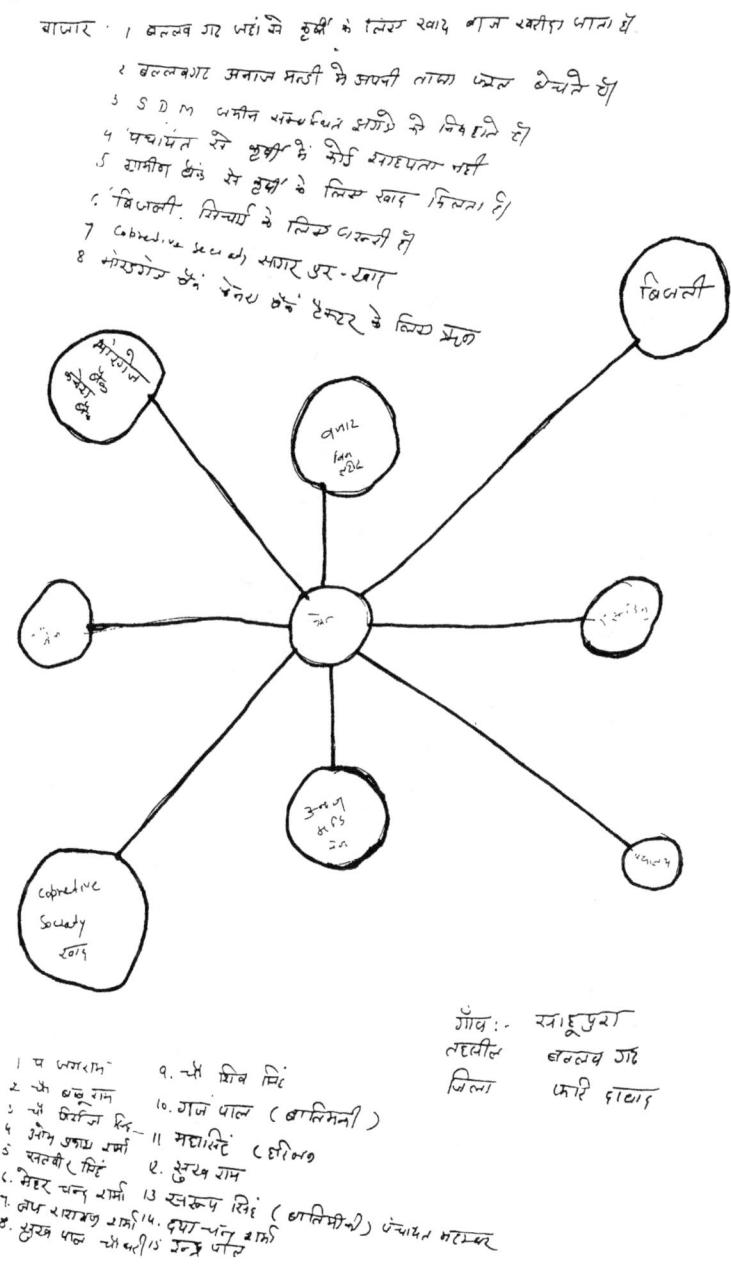

Chart 5a - Past Map
Village Chandawali, Faridabad

Chart 5b - Present Map
Village Chandawali, Faridabad

चन्दावली आज

2.

Chart 5c - Future Map
Village Chandawali, Faridabad

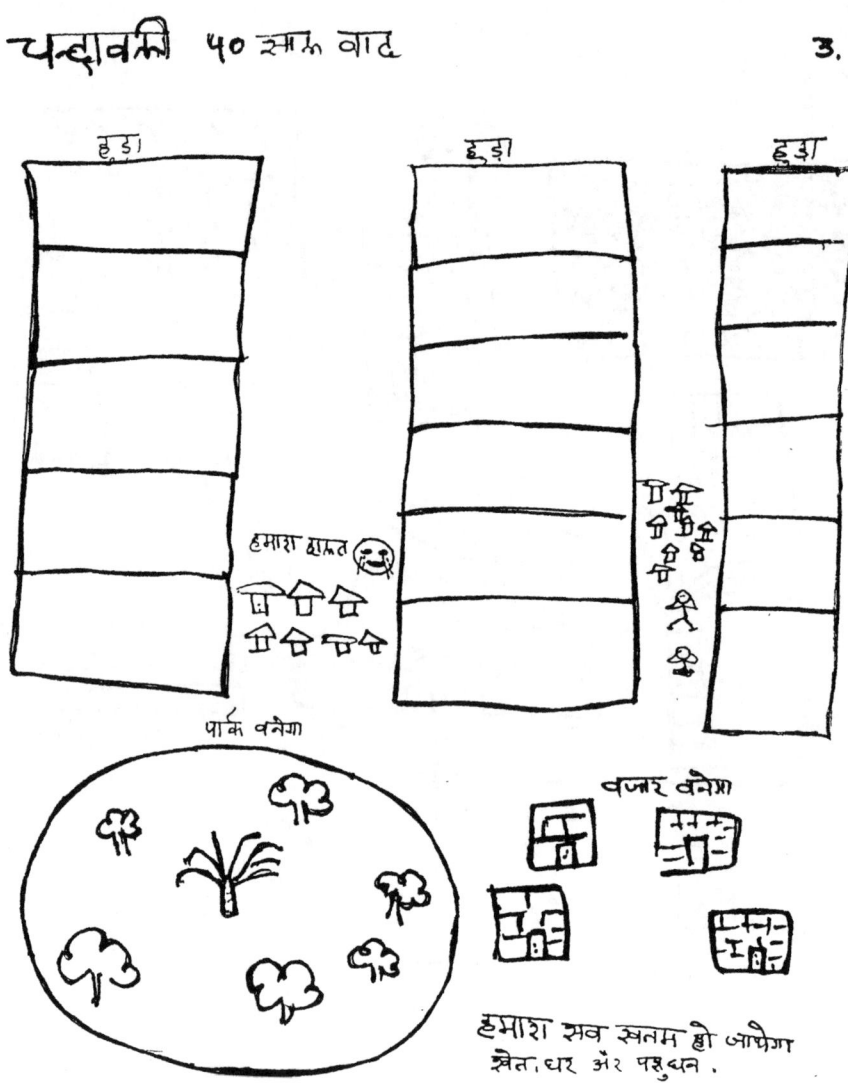

Appendix 2
Physical Maps of Selected Villages of Varanasi and Faridabad

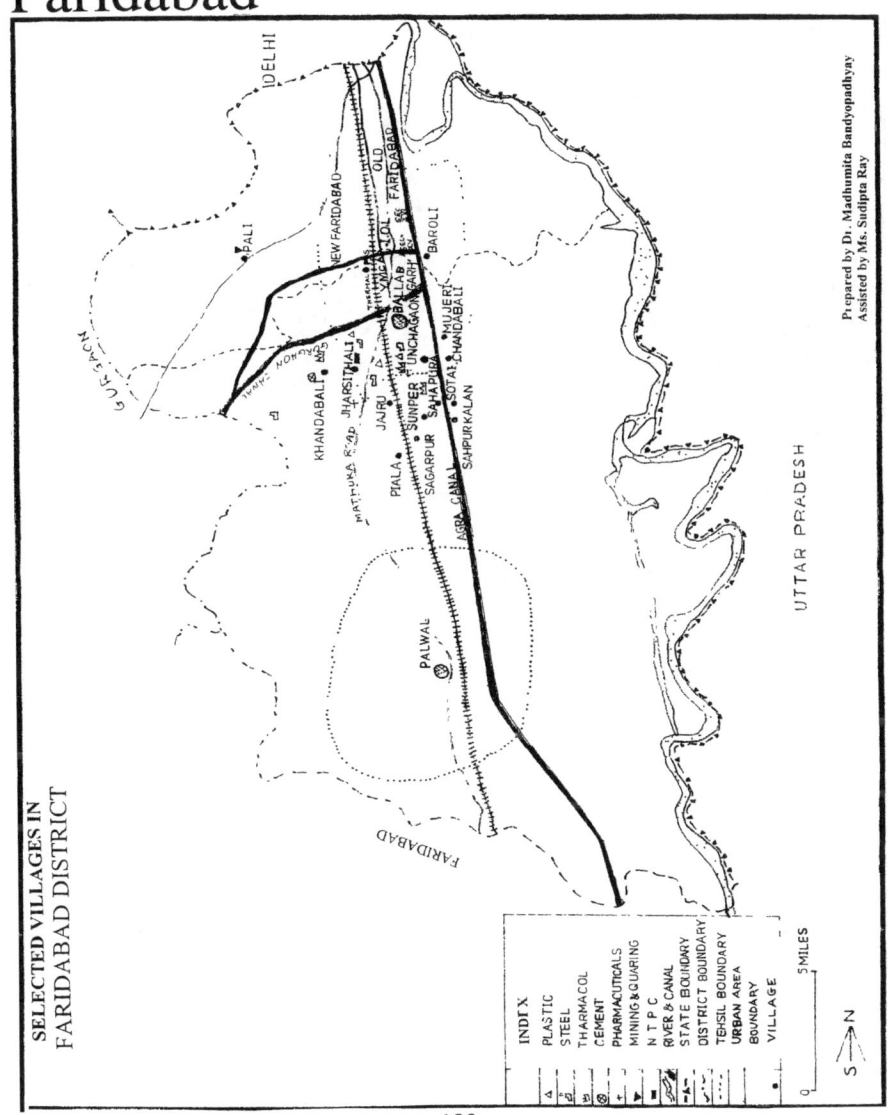

SELECTED VILLAGES IN FARIDABAD DISTRICT

Prepared by Dr. Madhumita Bandyopadhyay
Assisted by Ms. Sudipta Ray

UTTAR PRADESH

INDEX

◁	PLASTIC
☐	STEEL
♎	THARMACOL
⊗	CEMENT
+	PHARMACUTICALS
✦	MINING & QUARING
▆	N T P C
≈	RIVER & CANAL
— ‥ —	STATE BOUNDARY
·····	DISTRICT BOUNDARY
—·—	TEHSIL BOUNDARY
⬭	URBAN AREA
	BOUNDARY
•	VILLAGE

0   5 MILES

S ← N

199

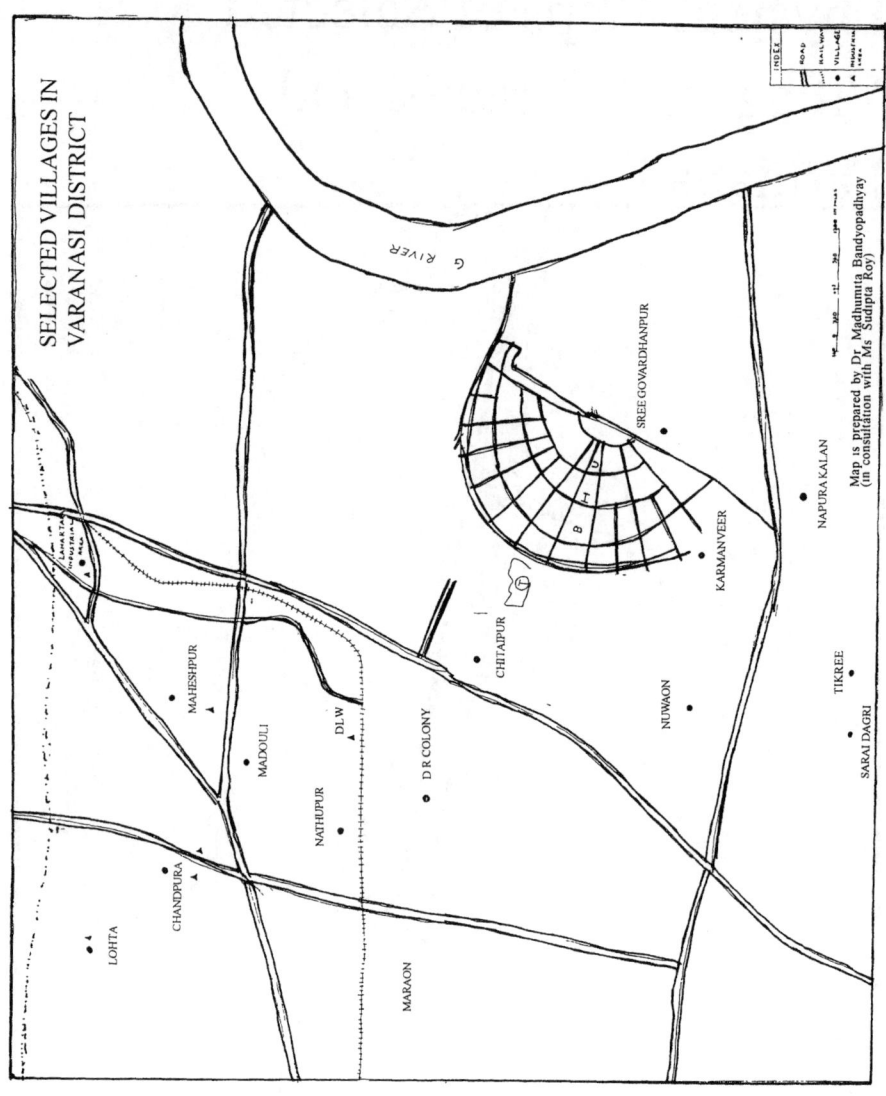

SELECTED VILLAGES IN
VARANASI DISTRICT

Map is prepared by Dr Madhumita Bandyopadhyay
(in consultation with Ms Sudipta Roy)

Bibliography

Collins, J. J. and D. Harris (1983), 'Air Pollution assessment at the International Crops Research Institute for the Semi-arid Tropics, India', *Environmental Technological Letters*, Vol..8. pp.10785-10787.

Conway, Gordon and Jules Pretty (1995), *Unwelcome Harvest*, Earthscan Publications Ltd. London and Vikas Publishing House, New-Delhi.

Halbwachs, G. (1984), 'Organisational Responses of Higher Plants to Atmospheric Pollutants: Sulphur Dioxide and Fluoride', in M. Treshow (ed.): *Air Pollution and Plant Life*, John Wiley and Sons, Chichester.

Khoshoo, T.N. (1984), *Environmental Concerns and Strategies*, India Environmental Society, New Delhi.

Markhan, Adams (1994), *A Brief History of Pollution*, Earthscan Publications Ltd., London.

Mukherjee, Neela (1994), *Participatory Rural Appraisal, Methodology and Applications*, Concept Publishing Company, New Delhi.

Mukherjee, Neela (1995), *Participatory Rural Appraisal & Questionnaire Survey: Comparative Field Experience & Methodological Innovations*, Concept Publishing Co., Delhi.

National Institute of Urban Affairs (2000), 'The role of urban and peri-urban agriculture in metropolitan city management in the developing countries', *Research Study Series, Number 74*, New Delhi.

Roberts, T.M (1984), *'Long Term Effects of Sulphur Dioxide on Crops: an analysis of dose-response relations'*, Phil. Trans. Royal Society London, No.305.

Rodhe, H. and R. Harrera (1988), *Acidification in Tropical Countries*, John Wiley and Sons, Chichester.

Runeckles, V.C. (1984), 'Impact of Air Pollutant Combinations on Plants', in M. Treshow (ed.), *Air Pollution and Plant Life*, John Wiley and Sons Chichester.

Simmons, I.G. (1993), *Interpreting Nature, Cultural Constructions of the Environment*, Routledge, London.

UNDP (1992), *Human Development Report*, Oxford University Press, New Delhi.

UNDP (1996), *Report on Human Development in Bangladesh*, A Pro-Poor Agenda, volume 3, United Nations Development Programme, Dhaka.

World Resources Institute (1998-99), *World Resources*, A Guide to the Global Environment, Oxford.

Index